STEPPING UP TO ISO 14000

STEPPING UP TO
ISO 14000

Integrating Environmental Quality
with ISO 9000 and TQM

Subhash C. Puri

Publisher's Message by Norman Bodek

Productivity Press • *Portland, Oregon*

Productivity Press
P.O. Box 13390
Portland, OR 97213-0390
United States of America
Telephone: 503-235-0600
Telefax: 503-235-0909
E-mail: service@ppress.com

Book design by Bill Stanton
Cover design by Gretchen Long
Page design and composition by Frank Loose Design
Printed and bound by Edwards Brothers in the United States of America

Library of Congress Cataloging-in-Publication Data

Puri, Subhash C.
      Stepping Up to ISO 14000: integrating environmental quality with ISO 9000 and TQM / Subhash C. Puri; publisher's message by Norman Bodek.
          p.   cm.
      Includes bibliographical references and index.
      ISBN 1-56327-129-X (hardcover)
        1. ISO 14000 Series Standards.   2. ISO 9000 Series Standards.
    3. Environmental management.   4. Total quality management.   I. Title
      TS155.7.P87   1996
      658.4'08--dc20   96-13221
                                         CIP

01 00 99 98 97 96 10 9 8 7 6 5 4 3 2 1

to my wife, **Shashi** and daughters,
Pamela and **Anuradha** for their patience,
understanding, and support

CONTENTS

PUBLISHER'S MESSAGE

A total reciprocal relationship exists between all living things in an environment and the environment itself. Any educated person today understands that the ocean is a complete ecosystem in which all the living things within the ocean are condensations of that environment, expressions of that environment. You have the ocean and then you have the fish swimming in it and the plants, and so on. Is the life that exists in the ocean separate from the ocean? Is that possible? The same thing holds true for human beings. We are walking around in an ocean. It has a little different humidity, a little different pressure to it. It is not as dense as the ocean atmosphere, but it's an atmosphere, nevertheless. And the quality of the environment exists in a reciprocal relationship to the quality of our life.

Swami Chetanananda

Quality programs in manufacturing companies that do not take into consideration environmental issues are only a small way down the road toward achieving excellence. Wherever organizations have considered the environment in which they produce and sell products as part of the system for which they are directly responsible, and have built programs to maintain or improve the quality of that environment, tremendous benefits to those companies have accrued in cost savings, customer satisfaction, and profitability.

The International Organization for Standardization (ISO) has been working diligently for a number of years to establish environmental quality standards for manufacturers. A number of countries throughout the industrial world have developed their own standards which are now being merged into the new ISO 14000 requirements. There is a great competitive advantage to be gained by those who adopt these new standards first, but there is no question that everyone will have to step up to them soon.

In *Stepping Up to ISO 14000: Integrating Environmental Quality with ISO 9000 and TQM*, Subhash Puri clarifies the fundamentals and advantages of environmental quality management. He details the steps and phases of implementing the new standards within an organization already committed to total quality management (TQM) and certified or seeking certification in ISO 9000. Dr. Puri also discusses critical issues in implementation for both established and new facilities and offers readers a road map for developing and implementing an environmental management system (EMS), including environmental performance evaluation, environmental risk assessment, and environmental quality auditing. A review of existing EMS standards in Britain, Canada, South Africa, and 12 other international EMS protocols are included as well. The new ISO 14000 specifications are explained and reviewed in relation to ISO 9000 requirements, including a checklist for implementation of ISO 14001 and one for integration of ISO 14001 with ISO 9001.

A summary of the author's approach to developing and implementing a TQM model is also presented in Appendix B. For details of this TQM model and its relation to ISO certification we refer our readers to Dr. Puri's first book, *ISO 9000 Certification and Total Quality Management*, also available from Productivity Press.

We are grateful to Dr. Puri for choosing Productivity Press to publish his fine book. There are thousands of books on ISO certification and until recently we have not chosen to publish on this subject. However, our readers have requested information that ties the ISO certification process into the continuous quality improvement model. This model is after all the foundation of our publishing strategy. We are delighted to have found Subhash Puri and take pleasure in extending his clear thinking on this subject to our readers.

In addition, we wish to thank all those who have participated in bringing this information from manuscript to bound book: Diane

Asay, editor in chief, for bringing Dr. Puri to us; Gary Peurasaari for editing the manuscript; Aurelia Navarro for copy editing; Pauline Sullivan for proofreading; Susan Swanson for managing the production process; Janet Brandt/Frank Loose Design for typesetting; Gretchen Long for cover design.

Norman Bodek
Publisher

There is a growing concern all over the world about the quality of our environment. This quality is largely dependent on the way we carry out our activities and utilize our materials, products, and manufacturing processes. We are already witnessing a tremendous growth in global industrial production, and there is a sense of uneasiness that the environmental burdens resulting from our uncontrolled manufacturing activities may be inflicting irreparable damage to our planet. The indicators of depleting environmental quality are conspicuously evident in many direct and indirect forms, such as the greenhouse effect, global warming, ozone depletion, loss of species and habitat, and air and water pollution.

If we do not take timely action to control our activities and their environmental impacts, we endanger the safety of our future generations. This issue is on the agenda of every concerned organization. Fortunately, efforts all around the world are already underway to address these concerns. International standards are being developed to assist organizations in controlling the environmental impacts of their activities; regulations are being proposed for the protection of the biosphere; and environmental management certification programs are being put into place, to name just a few. At the International Space University, we are attempting to make our contribution to the subject by holding international symposia and incorporating environmental aspects into our curricula in order to promote understanding and to study the use of space technology for environmental protection.

There are not too many publications available on the subject of environmental quality and safety, and perhaps none which provide simple guidelines for implementing environmental management systems. Subhash Puri, a leading authority on the subject of quality management, has once again come forward to share the fruits of his experiential knowledge. As with several of his other books on the subject, Dr. Puri lays out clear, easy-to-follow, step-by-step procedures that can be gainfully used by organizations seeking to implement environmental quality systems and prepare for certification to the international Environmental Management System (EMS) standard.

The author has earned an enviable reputation for his contribution to the appreciation and understanding of quality management through several books and professional papers over the last two decades. This book clearly manifests his ingenuity, comprehension, and command of the subject. *Stepping Up to ISO 14000* is a timely and valuable addition to the subject and the author should be congratulated and complimented on his dedication and contribution to the field of environmental quality.

Professor Ram Jakhu
Head, School of Management and Social Sciences
International Space University
Strasbourg, France

"Reduce - Reuse - Recycle - Recover"

Human capacity knows no bounds. We have the potential to create miracles. Over the past few centuries, we have achieved scientific and technological advancements never before envisaged. We are constantly endeavoring to enrich our lives and realize greater affluence. However, as intelligent beings, every once in a while we pause to reflect on our past accomplishments to make an objective appraisal of the strengths and weaknesses of our actions and achievements. We do this to ensure that we are on the right track and to take any action necessary to correct the situation, if required. Indeed, we do not wish to see the same technology that has helped us to grow and prosper become the instrument of our destruction.

This is the juncture at which we find ourselves today. We are getting signals that our technological achievements and their resultant outputs may have reached a critical stage and may be proving detrimental to the safe functioning of our natural ecosystem. There is a prevailing sense that our industrial infrastructure may be growing faster than the assimilation capacity and capability of the environmental ecosystem. Our present methods of expansion of the industrial systems seem to be lacking in a commitment towards either the protection of the biosphere or in the physical aspects of the availability of adequate managerial, technical, and financial resources to effectively address the environmental loadings resulting from these high levels of manufacturing activities.

Indeed, we cannot afford to ignore these issues, for what can be more important to us than the safety of our planet? If we do not have a habitable planet, then no amount of technological advancement

ievement has any meaning for us. We know, of course, that we
ı't let it happen. What we need to do is to control our activities
ıd their environmental impacts by modeling our systems on good
environmental management principles so as to bring harmony between the industrial and natural systems.

Environmental issues are now commanding global attention.
Virtually every organization, both in the public and private sectors, is
equally concerned about the matter and collective efforts are underway to address the growing problem of environmental safety. Many
concerned organizations have issued standards and guiding principles for environmental management as well as business charters for
sustainable development. While the regulators are considering establishing stringent rules and regulations for the protection of environmental quality, the manufacturing organizations are organizing
to voluntarily take responsibility for self-regulating their functions
through the implementation of effective Environmental Management Systems (EMS).

These concerns are also being collectively addressed at the international level through the establishment of the technical committee, ISO/TC 207, of the International Organization for
Standardization (ISO). The committee has been entrusted with the
responsibility of developing standards and guidelines to address all of
the important aspects of environmental management relating to
products, processes, and services. With this mandate, the committee
has already developed a set of draft standards under the ISO 14000
series. For example, ISO 14004 is an EMS standard that would provide guidance on implementing an environmental management system; ISO 14001 lays out core EMS specifications that will be used for
EMS certification, whenever the certification process is established;
ISO 14011 deals with environmental auditing; etc. This initial set of
standards is expected to be ready and available for use by the middle
of 1996. The standards in the ISO 14000 series share the same basic
system management principles found in the ISO 9000 series of quality system standards. Throughout this book I have used the draft
standards for the ISO 14000 series. While the ISO 14000 series

should be available to organizations by the middle of 1996, it is hoped that by the end of 1996, an integrated certification to both the ISO 9000 and ISO 14000 series will be available.

Since the subject of environmental quality management is relatively new, there are very few publications available on the subject. In this book, we are providing general guidelines on implementing an environmental management system and outlining the overall generic framework of EMS principles as appended in the forthcoming ISO 14000 series of EMS standards. For more details, the reader is advised to consult the standards when they become available.

The special feature of the book is its emphasis on establishing integrated quality systems: a Total Quality Management (TQM) system for the improvement of products and services; and an Environmental Management System (EMS) to control the environmental quality aspects of products and services. Therefore, guidelines are also provided for the certification of EMS and TQM systems to the respective ISO 14000 and ISO 9000 series of standards. Based on my extensive consultation experience with numerous international companies around the world, I would recommend that organizations that intend to implement and gain certification for both the EMS and TQM systems develop and establish a single integrated system rather than two noninteractive systems. This book contains an extensive set of guidelines, road maps, and checklists that an organization can utilize in their implementation of and certification to both the EMS and TQM systems. For readers who are not familiar with the TQM methodology, Appendix B gives a thorough introduction to its concepts and the steps for implementation.

I would like to express my gratitude to Professor Ram Jakhu for his continued support, and for writing the foreword to my book. Sincere thanks to my friends Roger Trudel and Om Kaura for their continued help, support and inspiration, and to Sylvie Olsen for her devoted assistance in the preparation of this book. Lastly, my utmost gratitude to my family for their love, care, patience and support.

Ottawa, Canada   Subhash C. Puri

PART 1

ENVIRONMENTAL QUALITY MANAGEMENT

1

THE QUALITY IMPERATIVE

THE PARADIGM OF SUCCESS

What is the most compelling drive in the business world? Undoubt-edly, it is *success*. Success translates itself into different configurations for different organization types. It is multifaceted and it requires mul-tidimensional activities. Organizations cannot simply produce prod-ucts and services the way they want, and expect to be successful. They have to satisfy the demands and expectations of various stakeholders—internal customers, external customers, regulatory agencies, and the societal infrastructure. They have to sustain profitability, credibility, and marketability or, if a public sector organization, gain public trust and confidence. And most importantly, they have to be cognizant of the fact that the single most crucial element in the equation of busi-ness successes and failures is *quality*.

As if this was not enough of a challenge for the business world, a new dimension has been added to the meaning of success. With the marketplace constantly undergoing dramatic transfor-mations, the connotation of "success" has been expanded to *com-petitive success*. There is no denying that with global markets shrinking, competition has become fiercely intense. Consequently, the success a company achieves can only be meaningfully evaluated in relation to its competition.

In searching for the determinants of competitive success, we know that to be competitively successful in the year 2000 and beyond, organizations must consistently provide products and services that, at a minimum, are technologically superior, possess quality excellence, are competitively priced, provide customer satisfaction, and are environmentally safe.

The Quality Factor

With the rapid growth of scientific knowledge and advanced technology around the world, most companies are capable of producing more volume, more variety, and cheaper goods. In a scenario such as this, the single most crucial element of competitive success is *quality*. When everything else is equal, quality provides the winning edge. But much like the paradigm of success, the scales of quality have also undergone dramatic shifts. The international standard, *ISO 8402: Quality Vocabulary*, aptly defines quality as follows:

> **Quality:** Totality of characteristics of an entity that bear on its ability to satisfy stated and implied needs.

Consequently, in today's competitive marketplace, the success paradigm does not revolve around *quality*, but *competitive quality*, i.e., an entity that collectively embodies such interactive processes as meeting specifications, exceeding expectations, consistency, innovativeness, and excellence.

To improve the quality of its products and services, a company has to establish, implement, and maintain an effective quality management system. The question that confronts most managers is which ready-made model would be most appropriate to their current operations. Many options for implementing a quality management system are available, such as:

- Develop customized model/system, based on the fundamental principles of quality and commensurate with current processes, procedures, and practices.

- Develop a model based on the guidelines outlined in the international standard, *ISO 9004-1: Quality Management and Quality System Elements—Guidelines*.

- Follow and implement the models, philosophies, and methodologies propounded by various quality experts, such as: Deming, Juran, Crosby, Fiegenbaum, and Ishikawa.

The Environmental Factor

There is growing public concern about the decreasing quality of our environment, and how our use of materials, products, processes, and services impact on it. Even though the quality of our products and services may be very good, their environmental impacts may be devastating. The total quality management (TQM) goal to serve customers has now been expanded to include the organization's social and environmental impact. This responsibility adds yet another dimension to the prevailing success paradigm—*environmental quality and safety*. After all, of what good to society is the success of an organization or the quality of its products and services, if these same products and services deplete the quality of the environment, directly or indirectly? Consequently, we can expect the new vision of *competitive* success in the year 2000 and beyond to require attention in almost equal proportion to quality management of products and services and to quality management of environmental impacts. Figure 1-1 shows the evolution of the success paradigm.

Environmental issues generally revolve around those activities that are associated with the life-cycle stages of an organization's products and services. These activities may include:

- Acquisition and utilization of raw materials

- Generation of processes

- Manufacture of material and products

- Provision of services

- Use of products, processes, and services, including maintenance, repair, reuse, and distribution

- Waste management, including recycling, disposal, and recovery processes

To set up an effective Environmental Management System (EMS) to improve the quality of the environment, the organization can build an appropriate system based on the guiding principles outlined in many environmental standards, legislative and regulatory requirements, or international protocols such as the *British Standard, BS 7750: Specification For Environmental Management Systems;* or *Canadian Standards Association, CSA-Z750-94: A Voluntary Environmental Management System.* (See Appendix A for a further listing of EMS protocols.) Or the organization can now develop an environmental quality model based on the guidelines outlined in the draft international standard, *ISO 14004: Environmental Management Systems—General Guidelines on Principles, Systems, and Supporting Techniques.*

ENVIRONMENTAL MANAGEMENT SYSTEM CERTIFICATION

Once an organization has implemented a quality system and has a reasonable degree of confidence in attaining high levels of product service quality, accreditation is usually sought to demonstrate that the company is operating under sound quality system principles. Traditionally, this was accomplished through two-party quality system audit programs with each customer, which were quite costly and time consuming. To overcome this wastefulness, most companies are seeking third-party certification for their quality systems against well known awards and standards, like the Malcolm Baldrige Award, Deming Prize, George M. Low Award, NASA's Quality and Excellence Award, Canadian Award for Business Excellence, and the Shingo Prize. More recently, the ISO 9000 Quality System Certification is commanding global acceptance and is being seriously sought by most organizations all over the world.

The focus on environmental quality management is relatively new. In the past, most organizations assumed that good manufacturing

Competitive Success in the Year 2000 and Beyond

Requires attention in almost equal proportion to:
Quality management of products/services
Quality management of environmental impacts
Success paradigm revolves around competitive quality

Environmental Quality and Safety

Identifying the organization's social impact and
serving the broader customer environment
Quality management of environmental impacts

Competitive Quality

Embodies interactive processes of—
meeting specifications
exceeding expectations
consistency
innovativeness
excellence
Quality management of products/services

Competitive Success

technologically superior
quality excellence
competitively priced
customer satisfaction
environmentally safe

Old Success Model

profitability, credibility
marketability, quality,
satisfy demands of regulatory
agencies and societal infrastructure

Figure 1-1. Evolution of the Success Paradigm

practices adequately addressed environmental concerns. There were no well-structured systems or methodologies to control the environmental impacts of products and services. Organizations simply identified the environmental aspects of their activities and attempted to minimize the environmental burdens through the application of sound environmental quality management principles or compliance with the requisite legislative and regulatory requirements. Consequently, the demonstration of a company's commitment to environmental quality and safety was not by means of any structured certification system but through the process of self-regulation.

As concerns about environmental safety and protection grew over time, several national and international standards and protocols came into being. One early standard that gained reasonably wide acceptance was the *British Standard, BS 7750: Specification For Environmental Management Systems*. In fact, quality system registrars in several countries are providing third-party certification services to companies against the EMS specifications of BS 7750.

Today, with environmental issues continuing to command global attention, the International Organization for Standardization (ISO), through its technical committee, TC 207, has developed the following documents:

- ISO 14004: Environmental Management Systems— General Guidelines on Principles, Systems, and Supporting Techniques

- ISO 14001: Environmental Management Systems— Specifications with Guidance for Use

While ISO 14004 will provide guidance for establishing an EMS, ISO 14001 will be used for EMS certification. A third-party certification process (much like the ISO 9000 process) is being established by the ISO for EMS certification against the requirements and specifications as outlined in ISO 14001. In Chapters 6 and 7 we discuss the process of certification to ISO 14001.

QUALITY MANAGEMENT SYSTEMS

Total Quality Management (TQM)

One of the fundamental aims of any manufacturing organization is to improve the quality of its products and services. Traditionally, this was accomplished through the application of product oriented methods of *quality control* and *quality assurance*. Today, the emphasis is on establishing an integrated whole system—a total quality management system (TQM)—to collectively improve all aspects of operational and functional efficiency.

What is TQM? It is simply a set of system elements which when implemented properly, bring regimentation, systematization, and discipline into the routine operational and functional framework of the company, by virtue of which quality of products and services can be achieved, maintained, and improved. (To implement, establish and maintain a total quality management system using a step-by-step procedure, see Appendix B.)

As a sum total of system elements, quality activities and initiatives, a TQM system can be implemented through the systematic establishment and maintenance of the following basic elements:

- **Management:** vision, mission, commitment, leadership, quality policy and objectives

- **Strategic planning:** quality plans, procedures, processes, activities, quality infrastructure

- **Human Resource Management:** employee involvement and empowerment, process improvement teams, training and development

- **Input Controls:** market analysis, customer needs, incoming quality assurance, supplier-customer partnering

- **In-Process Controls:** control of design, specifications, material, equipment; process control; documentation control; production control; process/product audit and verification

- **Measurement and Analysis:** data collection and analysis, statistical process control

- **Output Controls:** output audit and verification; conformance to contractual and regulatory requirements

- **Customer Satisfaction:** customer requirements and expectations, services, standards, commitment, complaint resolution, satisfaction determination, satisfaction results

It should be clearly understood that all the models, whether for quality system implementation or certification, are built on a common core set of quality principles. The differences, if any, lie in their focus, approach, and philosophical slant. Although any model can provide suitable quality system framework, success will be determined by the organization's ability to manage and control the system effectively. The greatest amount of lasting success will be achieved by those organizations who develop their own quality model commensurate with their specific needs, practices, and infrastructural profile.

It must be stressed that no system is capable of turning things around overnight; quality cannot be instantly manufactured. It is infused and imbedded into a product or service through systematic means. Japan's quality superiority did not spring up overnight. Quality is achieved, piece by piece, process by process, under the overall umbrella of a total quality management system. Quality is the end result of proper planning, patience, dedication, and interactive and integrated systems. Quality is both a short-term function and a long-term focus.

Environmental Management System (EMS)

The core principles of a quality system are the same whether the system is for the quality management of products and services or for the management of environmental quality. The differences arise in the nature of the entity, its specific requirements, and the scope of its applications. To develop a suitable quality management system for environmental issues, we need to reengineer the core principles of quality management to address the specific nature and requirements of the issues relating to the environment.

This means that in order to establish an environmental management system (EMS), an organization must identify the nature of its activities, products, processes, and services, as well as the environmental aspects and impact of its activities. The organization also needs to be aware of the environmental concerns of all the stakeholders and identify any pertinent legal and regulatory requirements.

Once this package of information is ready, the framework for the EMS can be easily established. The following sequence can be considered for establishing the environmental quality system:

- Establish an environmental policy. The policy must accommodate the concerns of all stakeholders and all requisite legal and regulatory requirements.

- Establish objectives and targets for the environmental aspects and impact of activities.

- Delineate appropriate responsibilities and allocate resources.

- Establish environment management initiatives, projects, and programs.

- Establish operational controls.

- Measure and monitor system effectiveness.

- Establish continual improvement processes and procedures.

The various approaches to establishing and implementing an EMS are outlined in the succeeding chapters.

2

ENVIRONMENTAL QUALITY
MANAGEMENT FUNDAMENTALS

PROTECTION OF THE BIOSPHERE

Worldwide public concern about the quality of our environment is growing. With the industrial infrastructure growing faster than the assimilation capacity and capability of the environmental ecosystem, the concern for the protection of the environment and, consequently, of human health and safety, is reaching a critical point. As a result, environmental issues are now commanding global attention. What constitutes the environment? The draft international standard ISO 14001 defines it as follows:

> *Environment:* Surroundings in which an organization oper-
> ates, including air, water, land, natural resources, flora,
> fauna, humans, and their interrelation. The environment
> in this context extends from within an organization to the
> global system.

The quality of the environment is dramatically affected by the way we utilize materials, products, and energy, and by our manufacturing processes and their by-products. Most products, processes, and services follow a life-cycle pattern, and depending on their nature and complexity, each stage of their life-cycle has some impact on the environment. The loadings on the environment can be in the form of gaseous emissions to the air or liquid or solid wastes discharged to the soil or water. The failure to effectively control these entities is resulting in the loss of natural resources, loss of species and habitat, decreased biodiversity, degradation of water and air quality, and dumps and landfills. Furthermore, there are the many direct and indirect forms affecting the biosphere, such as the greenhouse effect (global warming) and ozone depletion.

What actions are required to address these concerns? As a starting point, an organization needs to initiate the following:

- A set of guidelines for the manufacturing organizations to develop and implement an effective environmental quality management system.

- A protocol for sustainable development. Sustainable development can be defined as: "Development that meets the needs of the present without compromising the ability of future generations to meet their own needs."

- Formal regulatory and legislative framework for industry compliance and monitoring to environmental criteria.

- An internationally accepted set of specifications for establishing an effective Environmental Management System (EMS).

- An internationally accepted certification process, like the ISO 9000 certification process, through which the organizations can seek accreditation for their environmental systems and demonstrate compliance with the accepted norms and specifications.

- An environmental audit system.

- A tangible commitment from the manufacturing and service sectors to self-regulate their functions.

Indeed, most of the current efforts are already being expended along these lines. For instance, many concerned organizations have issued standards and guiding principles for environmental management as well as business charters for sustainable development. A few examples of such standards and protocols are as follows.

EMS Standards

- British Standard, BS 7750: Specification for Environmental Management Systems

- Canadian Standards Association, CSAZ750-94: A voluntary Environmental Management System

- South African Standard, SABS-0251: Environmental Management System

EMS Protocols

- The Rio Declaration on Environment and Development: The United Nations

- International Chamber of Commerce (ICC): Business Charter for Sustainable Development

- European Petroleum Industry Association (EUROPIA): Environmental Guiding Principles

- Keidanren (Japan Federation of Economic Organizations): Keidanren Global Environment Charter

- Coalition for Environmentally Responsible Economies (CERES) Principles (formerly the Valdez Principles)

- Business Council on National Issues: Business Principles for a Sustainable and Competitive Future

(cont'd)

- National Round Table on the Environment and the Economy (NRTEE): Objectives for Sustainable Development

- Canadian Chemical Producers' Association (CCPA): Responsible Care—Guiding Principles

- The Environmental Policy of the Coca-Cola Company

- The Environmental Policy of British Telecom

- The Environmental Policy of National Power

- World Industry Council for the Environment (WICE): Guidelines on Environmental Reporting

INTERNATIONAL ORGANIZATION FOR STANDARDIZATION (ISO)

The international response to the environmental concerns is being addressed through the technical committee ISO/TC 207 of the International Organization for Standardization (ISO). The committee has been entrusted with the task of developing standards and guidelines to address all important aspects of environmental management relating to products, processes and services. Most of these standards fall under the ISO 14000 series.

While the standards in the current ISO 9000 series address the issues of quality assurance and quality management of products and services, the standards in the ISO 14000 series address the environmental aspects of those products and services. However, both series share common system management principles.

The technical committee, ISO/TC 207, has been mandated to develop all relevant guidance and compliance standards in the ISO 14000 series that would appropriately address the requisite environmental issues and concerns. As a first attempt at the series, the committee developed a set of draft standards as summarized in Table 2-1. Also, in its continuing effort to develop standards on supporting environmental technologies, the committee is working towards developing additional standards in the next few years, as detailed in Table 2-2.

The most important of the initial set of standards are ISO 14001 and ISO 14004. ISO 14001 is the standard that specifies the core requirements for environmental management systems, and is intended to be used as a conformance standard for EMS certification, wherever the certification process is established. ISO 14001 shares common system management principles with ISO 9001. The other standard, ISO 14004, is a guidance standard addressing a broad range of environmental management issues. It provides simple guidelines for establishing and implementing an EMS. In the ISO 14000 series, this standard plays the same role that standard ISO 9004-1 does in the ISO 9000 series.

Another useful standard in the series is ISO 14011/1, which provides guidelines on environmental auditing procedures. This standard bears a common linkage to the quality system auditing procedures outlined in the standard ISO 10011/1.

In the next four chapters, we will outline the general framework of some of these important EMS standards to provide a preview of the forthcoming environmental quality system requirements that most organizations may have to comply with. In addition, simple guidelines are presented for implementing an environmental quality system and its certification to ISO 14001. Organizations can gainfully utilize the many checklists and road maps provided in this book to prepare their environmental systems in advance, for the forthcoming EMS certification specifications.

It should be clearly understood that during the developmental phase of international standards, changes or modification to either

Table 2-1. ISO 14000 Series—Environmental Standards

ISO 14001	Environmental Management Systems - Specifications with Guidance for Use
ISO 14004	Environmental Management Systems—General Guidelines on Principles, Systems and Supporting Techniques
ISO 14010, 14011/1, 14012	Guidelines for Environmental Auditing
ISO 14010	General Principles of Environmental Auditing
ISO 14011/1	Audit Procedures—Part 1: Auditing of Environmental Management Systems
ISO 14012	Qualification Criteria for Environmental Auditors
ISO 14024	Environmental Labelling—Practitioner Programs—Guiding Principles, Practices, and Certification Procedures of Multiple Criteria (Type 1) Programs
ISO 14031	Environmental Performance Evaluation
ISO 14040	Life-Cycle Assessment—Principles and Guidelines
ISO 14050	Terms and Definitions

Table 2-2. ISO 14000 Series—Environmental Standards (Future Work Items)

ISO 14013	Management of Environmental Audit Programs
ISO 14014	Initial Reviews
ISO 14015	Environmental Site Assessments
ISO xxxxx	Environmental Labelling
ISO 14021	Self-Declaration Environmental Claims— Terms and Definitions
ISO 14022	Symbols
ISO 14023	Testing and Verification Methodologies
ISO 14024	Practitioners Programs - Guiding Principles, Practices and Certification Procedures of Multiple Criteria (Type 1) Programs
ISO 14025	Goals and Principles of All Environmental Labelling
ISO 14xxx	Type III Labelling
ISO xxxxx	Life-Cycle Assessment
ISO 14040	Principles and Guidelines
ISO 14041	Life-Cycle Inventory Analysis
ISO 14042	Impact Assessment
ISO 14043	Interpretation
ISO 14060	Guide for the Inclusion of Environmental Aspects in Product Standards

the numbering system or headings of standards or to their content can be expected to occur. The reader is, therefore, advised to consult the approved standards as and when they are available for use.

ENVIRONMENTAL CONCEPTS

We are already witnessing a tremendous growth in global industrial production, and there is increasing concern that the environmental loadings resulting from these high levels of manufacturing activity may endanger the natural ecosystem. For example, we know that the extent of industrial flows of nitrogen and sulfur is already equal to or greater than the natural flow; and that the consumption of energy in the form of hydrocarbons is leading to the carbon dioxide greenhouse effect.

What is the greenhouse effect? As a natural phenomenon, the energy received from the sun as visible sunlight heats the earth's surface and is reflected back to space. Although some of the heat escapes back to space, much of it is trapped by the so-called greenhouse gases, creating additional heat to the earth and causing what is referred to as global warming. The major naturally occurring greenhouse gases are carbon dioxide (CO_2); water vapor (H_2O); methane gases (CH_4); nitrous oxide (N_2O); and ozone (O_3). But the expansion of industrial activities worldwide is elevating the levels of greenhouse gases in the atmosphere. For example, excessive burning of coal, gas and oil and depletion of forests are causing increasing levels of carbon dioxide. These increasing levels of greenhouse gases are further trapping the sun's energy and, therefore, raising the earth's surface temperature. Likewise, these natural and man-made gases are contributing towards the depletion of ozone levels, the consequences of which can damage agricultural systems and human health.

Our present industrial activities show neither a commitment towards the protection of the biosphere nor the availability of adequate managerial, technical, and financial resources to attack our growing environmental problems. It is imperative that organizations around the world model their industrial activities on good environmental management principles and make a serious commitment to effectively manage these activities so we can bring them in harmony with the natural environmental systems. If we do not take timely action to control these industrial activities and implement an effective environmental management system, we endanger the safety of future generations.

This goal can be accomplished through the establishment and maintenance of suitable environmental management systems. However, to develop such a system, it is imperative to clearly understand the nature and complexity of the problem. Below are some definitions of key words associated with the EMS. They are taken from the draft international standard ISO 14001.

> *Environmental Aspect:* Elements of an organization's activities, products, or services which can interact with the environment.

> **Environmental Impact:** Any change to the environment, whether adverse or beneficial, wholly or partially resulting from an organization's activities, products, or services.

> **Environmental Management System (EMS):** Organizational structure, responsibilities, practices, procedures, processes, and resources for implementing and maintaining environmental management.

Designing for the Environment

Products, processes and services of an organization exert environmental burdens at all stages of their life-cycle. If the activities that impinge on the environment can be designed properly in the first place, there can be significant source reduction of the environmental impact of these activities. A well designed environmental management system would have a high probability of success and sustainability.

Traditionally, organizations have been operating in a reactive mode, i.e., identifying the environmental aspects of activities and their impacts on the environment, and establishing an environmental management system to control those activities. Environmental compliance and proactive environmental consideration in activities such as the design function have rarely been considered. However, it is now being recognized that the designers are the ones who really understand the origins of products and how they impact the socioeconomic infrastructure of the society and, therefore, may be able to help evaluate real environmental performance of activities and propose genuine solutions to environmental concerns. In this way, instead of a piecemeal crisis-oriented response to environmental issues, the organization can incorporate all requisite design aspects into the system and synthesize the improvement initiative to meet their objectives.

Many useful benefits can accrue from *designing for the environment.* The organization can anticipate regulatory compliance requirements and incorporate them into the design of activities, products, processes, or services. It can properly design products that will yield efficiency and cost-effectiveness as well as reduce risk and liabilities. The organization can also achieve competitive advantages through better designs and enhance its capability to use the best available practices and state-of-the-art technology.

LIFE-CYCLE ASSESSMENT—ESTABLISHED FACILITIES

All environmental management systems are designed to ensure that the environmental consequences of an organization's activities are adequately controlled. For the purpose of environmental management, controls need to be established at two levels of activities: *established facilities* and *new facilities.* New facilities will be discussed in the next section.

For an already functioning manufacturing unit, effective controls need to be considered at the following two levels of activity:

- **Design Level:** to ensure that most of the activities are designed properly in the first place, so as to minimize their impending environmental impacts.

- **Operational Level:** to ensure that adequate procedures and practices are in place to improve and maintain the quality of the environment.

Since the aim in implementing an EMS is to control the environmental impact of activities, the most appropriate thing to do is to evaluate the life-cycle aspects of activities, identify their environmental effects and impacts, and then develop a suitable system that would adequately address the environmental issues.

The *life-cycle approach* is defined in *ISO 14031—Guidelines on Environmental Performance Evaluation*, as follows:

> **Life-Cycle Approach:** a tool which may embody analysis of possible environmental impacts as a result of raw material inputs to production through ultimate use and disposal of a product. The life-cycle approach may not necessarily utilize full scientific technology but will require making reasonable attempts to understand environmental relationship at each phase in the cycle of function or activity.

Most manufacturing situations involve six life-cycle stages (see Figure 2-1).

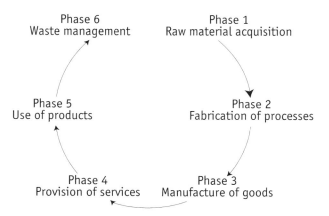

Figure 2-1. Six Life-Cycle Stages of Product/Services

For each of these life-cycle stages it is essential to consider two aspects:

- *Design elements* that can be considered and incorporated during the initial phases of production so as to control the potential life-cycle environmental impacts of products.

- *Operational elements*, including the extent of controls to be exercised during the production process.

Each life-cycle stage needs to be evaluated one by one in order to develop a profile of improvement practices that can be incorporated at the *design level* or *operational level*. The most appropriate approach to properly control the environmental aspects of activities for improving and maintaining environmental quality is to:

- Evaluate each life-cycle stage of activities

- Identify the environmental aspects of activities

- Assess the environmental impact of activities

- Establish critical control points within each life-cycle stage to monitor and control the environmental aspects of processes

- Establish a suitable overall environmental management system for the entire life-cycle framework of activities that will effectively manage and improve the quality of the environment

Span of Activities in the Life-Cycle Stages

The span of activities encompassed within the six life-cycle stages can be defined as follows:

1. *Raw Material Acquisition:* Materials acquired, extracted, derived, or harvested for use in the manufacture of products

and processes; quantities of materials; testing and sampling; transportation; the type, nature, and intensity of materials; material substitution and reformulation; discharge loadings and quality control.

2. *Fabrication of Processes:* Processes involved in the manufacture of goods that may have an environmental impact: chemical, physical, biological; water and energy consumption; discharges to air, water, and solid wastes.

3. *Manufacture of Goods:* Material management; material manufacture; material handling, storage, and transportation; product fabrication; product packaging and labeling; product delivery and distribution.

4. *Provision of Services:* Materials, methods, or processes involved in the provision of tangible or intangible services which may have an impact on the environment.

5. *Use of Products:* Removal and disposal of packaging; reuse, conversion, remanufacture; maintenance and repair; consumer use and misuse.

6. *Waste Management:* Recycling; recovery; composting; land treatment; land reclamation; landfill; incineration; refuse and disposal.

For the span of activities identified in each of these life-cycle stages, we will describe in Table 2-3 how to make an appraisal of their impact, what monitoring and control procedures to use to reduce or eliminate environmental loadings, and what improvement practices and management principles to employ for the protection of the biosphere.

Table 2-3. Life-Cycle Stages—Appraisal, Monitoring and Control, Improvement Practices

Stage 1: Raw Material Acquisition

- Make a checklist of all the different types of raw materials acquired from the suppliers and subcontractors for use in the processes leading to the manufacture of goods.

- Identify the environmental aspects and impacts of the raw materials.

- Ensure that the suppliers provide full specifications and details of raw materials and their environmental profile.

- Partner with suppliers to ensure that quality control procedures are appropriately maintained on all raw materials and that the materials conform to the requisite specifications.

- Make sure that your subcontractors clearly understand your commitment to environmental quality.

- Try to eliminate or reduce the use of material or energy that is not environmentally safe. Work with material substitution or reformulation and ensure that the raw material is extracted by processes and technologies that are environmentally clean.

- Use environmentally benign raw materials as far as possible. Simplify raw materials selection to limit the combinations of raw materials used.

Stage 2: Fabrication of Processes

- Identify all processes, formulations, and chemicals used for the manufacture of goods.

- Identify processes that discharge pollutants to air, water, and land.

- Evaluate the environmental impact of all these processes.

- Design materials and processes to be recyclable.

- Minimize or eliminate the use of environmentally unsafe materials and processes. Design with process substitution, process control, improved process layout, treatment, and disposal.

- In the case of chemicals or formulations that have a limited shelf life, acquire only minimal or requisite quantities.

- Identify if the process is necessary or if it can be subcontracted and fabricated under controlled conditions.

Stage 3: Manufacture of Goods

- Implement effective material management procedures.

- Identify all environmental burdens in the production, use, and disposal of materials.

- Use recycled, renewable, or virgin material.

- Use material available in bulk or in reusable and returnable containers.

- Buy materials marked with Eco label logos.

- For all acquired materials and components, ensure that they comply with requisite specifications with regard to: recycled content, reformulation, additives, etc.

- Establish suitable material handling, storage, and transportation procedures, including:
  – procedures for palletizing
  – procedures for packing or unpacking materials for storage

(Cont'd)

    – controlled environmental conditions for storage

    – effective storage and retrieval system

    – special procedures for materials with shelf-life considerations

    – reuse, and recycling aspects of materials

- In the manufacture of products, ensure that:
  - energy consumption levels are low or optimum
  - products are safe and fit for human use
  - there are no harmful by-products or side-effects from usage
  - products do not contain any harmful substances
  - it permits minimum raw material waste
  - it utilizes processes that are environmentally controlled

- Establish a profile of the manufacturing process, identifying:
  - process inputs: material, components, energy and water consumption, process transformations to materials
  - process outputs: products, co-products, contaminants, wastes

- Establish good environmental management practices to reduce environmental burdens by:
  - establishing energy efficient processes
  - complete elimination of the use of material or processes that are environmentally harmful
  - substituting or reusing a product or material in its original

form for applications similar to that for which it was originally intended

– developing material efficiency procedures

– developing optimum designs for the layout of facilities and processes

– establishing efficient treatment and disposal methods

– achieving source reduction of waste materials within the production process through a change in design, specification, or process

– establishing efficient recovery procedures, such as: composting, energy recovery, incineration, pyrolysis, etc.

– establishing good waste management practices: recycling, degradability, recovery and landfill

– developing viable design strategies for product life extension: adaptability, durability, maintainability, operating efficiencies, reliability, and reuse

• Establish guidelines for packaging in conformity with environmental quality and safety requirements. Some essential aspects include:

– packaging to consider aspects of source reduction, reuse, and recycling

– packaging to conform to health, safety, and sanitation requirements

– packaging should not pose any physical danger to the intended users

– viable design strategies to ensure that packages are functional, returnable, reusable, recyclable, and environmentally disposable

(Cont'd)
- packaging to conform to all requisite labeling requirements and regulations

- Establish controls for the process of delivery and distribution of products from the manufacturing location to the consumer location. Some essential aspects to be considered include:
  - Can containers be reused or recycled?
  - Are there special precautions that need to be taken for shipment of hazardous material?
  - Does the material emit toxins?
  - Are transport methods energy efficient?
  - Can shipment schedules be optimized?
  - Are shipment mediums equipped for emergency situations?
  - Has an emergency plan been developed in case of an environmental emergency?

Stage 4: Provision of Services

- Identify any aspects of provision of tangible or intangible services that may, directly or indirectly, impact the environment.

- Establish suitable monitoring and control procedures for the management of services to ensure that the environmental concerns are adequately addressed.

- Environmental aspects of services may include such entities as: design for new sites; packaging; material or product handling; storage and transportation; maintenance; waste disposal; and landfills. Attempts should be made to control the environmental aspects of services at the design stage, if possible.

Stage 5: Use of Products

The scope of this life-cycle stage extends from the removal and disposal of packaging, putting the product to service, operation and maintenance, until its final disposal. Some important environmental considerations in this stage of activity are:

- To ensure that products or materials are designed properly to be durable, reliable, adaptable, maintainable, reusable, recyclable, and energy efficient.

- If required, technological changes or readjustments should be considered to ensure that the product or material:
  - is functionally suitable

  - can be easily adapted to a multiplicity of purposes

  - can be reused and easily recycled

  - poses minimum environmental risks and burdens

- In its operational mode, the product should:
  - be functionally adequate and meet all environmental requirements and specifications

  - not pose any environmental loads, such as emission to air, water, or land

  - not pose any danger to its intended user

  - be reliable and operationally viable for the intended product life

(Cont'd)

- The product should be designed in such a way that its maintenance is easy and the maintenance materials and parts can be reused and recycled.

- The organization should ensure that while the product adequately fulfills its intended function, it does not lend itself to misuse through packaging, storage, unintended multiple usage, specification change, reuse, recycling, or any other means associated with its operation, maintenance, and disposal.

- If the disposal of the product involves any environmental burdens and risks, the manufacturer should ensure that these aspects are adequately accommodated and controlled either through design, manufacture, or information regarding the product or its disposal.

Stage 6: Waste Management

The disposal stage of the product or material life-cycle occurs when the product has served its intended purpose of design, manufacture, operation, maintenance, and use. Since a large majority of environmental burdens are associated with waste management, it is mandatory for the manufacturer to ensure that the product is designed and manufactured with due considerations to its disposal aspects, such as recycling, composting, land treatment, landfill, land reclamation, incineration, recovery, pyrolysis, and refuse.

Preparing an Environmental Profile for EMS

To summarize the life-cycle assessment, we will outline the sequence for preparing an environmental profile of activities to establish an effective EMS for monitoring, controlling, and improving the quality of the environment.

1. Develop a profile of the organization's activities.

2. Identify the nature and complexity of activities.

3. Identify the types of products fabricated.

4. Identify the life-cycle stages of activities.

5. Identify the environmental aspects of activities within each of the life-cycle stages.

6. Develop a profile of direct and indirect environmental effects of activities. Effects identification must be all-encompassing with respect to operating conditions: normal, abnormal, incidents, accidents, potential emergencies, etc.

7. Establish links between activities and environmental effects, e.g., emissions to air; discharges to water; solids and other wastes; contamination of land; use of land, water, fuels, energy, and other natural resources; effects of noise, odor, dust, vibration, and visual impact; effects on the ecosystem.

8. Develop an environmental policy, mission, and core values.

9. Specify inputs and outputs for the activities.

10. Identify stakeholders concerns.

11. Evaluate current systems, procedures, and processes in terms of their suitability to control environmental conditions.

12. Assess environmental impacts against current practices.

13. Identify gaps and deficiencies.

14. Develop an integrated quality-environment management system.

15. Formalize commitment and involvement.

16. Develop an EMS implementation framework.

17. Establish requisite environmental management procedures.

18. Assign appropriate responsibilities for the environmental management aspects and allocate appropriate technical, human, physical, or financial resources.

19. Establish process management and operational control activities.

20. Establish training and development programs.

21. Establish monitoring and measuring systems.

22. Establish procedures for environmental auditing and system performance.

23. Institute a continuous improvement program.

INTEGRATED EMS PROCEDURE—NEW FACILITIES

As mentioned earlier, there are two levels of activities in which controls need to be established for environmental management. We have discussed the effective controls needed for an established facility. In this section we will outline the proposal development procedure for activities associated with a new facility, and provide a checklist of environmental characteristics to be considered during the proposal development phase.

Whenever a new manufacturing facility is to be established, it would be a good practice to follow some EMS guidelines for developing the new site proposal so as to minimize environmental loadings and concerns right from the start. These proposals should address all pertinent environmental characteristics associated with the development and functioning of the new facility.

Proposal Development Procedure

- Establish the proposal framework: objective, scope, purpose, magnitude and complexity.

- Identify the nature and extent of the manufacturing activities to be carried out in the new facility.

- Identify the environmental aspects of the activities.

- Assess the environmental impact of the activities.

- Classify the proposal as *significant* or *not significant* in terms of the environmental impact of activities.

- Establish a detailed profile of the impact.

- Notify and seek input from all concerned parties.

- Establish environmental policy, objectives, targets, and core values.

- Identify legal and administrative requirements.

- Review or revise the proposal as required.

- Establish and record conditions of approval.

- Establish an implementation framework.

- Establish monitoring and auditing procedures.

- Outline procedures for activity review.

- Establish appeal procedures.

- Establish a mandate for amendment and approval.

Environmental Characteristics: A Checklist

Figure 2-2 outlines the major environmental characteristics which may potentially be affected by development actions, or which could place significant constraints on a proposed development. All of these entities and their environmental aspects must be carefully evaluated during the developmental phase of a new project proposal. Following is a generic checklist of these major environmental characteristics.

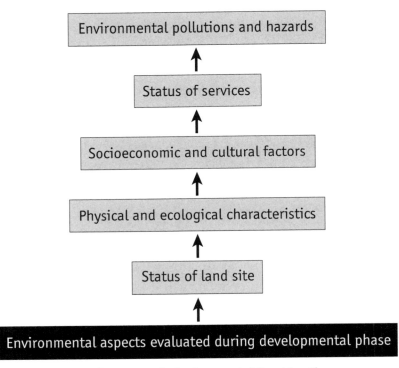

Figure 2-2. New Site Proposal—Environmental Considerations

Status of Land Site

Identify if the proposal development would impose a significant burden on any of the following:

General characteristics of the site: overall compatibility of land site with its intended use; compatibility with the scale of development, building codes, and other socioeconomic, political, and legal aspects; aesthetic characteristics of the landscape; natural ecosystem.

Residential areas: compatibility with the surrounding residential developments; community infrastructure, services, lifestyle, and core values.

Commercial areas: compatibility with other existing commercial entities; nature and characteristic of commercial activity, provision of services, facilities, and modus operandi of people.

Industrial areas: compatibility with environmental burdens of existing industrial infrastructure, levels of pollution, waste management activities, and provision of services.

Agricultural areas: compatibility with the natural ecosystem, farmland environment, greenbelt obligations, pollution levels, disease control, soil erosion, plant and animal life, and services.

Physical and Ecological Characteristics

Assess the impact of the proposed development vis-à-vis the following factors:

Physical characteristics of the land: its surface, nature of substrata, seismic activity, water logging aspects, soil and sedimentation aspects, and its general stability.

Climatic conditions of the area: wind strength, rainfall patterns, temperature and humidity aspects, etc.

Conditions relating to vegetation, animals, marine and freshwater systems: natural balance of plant and animal life; endangered plant and animal species; conservation concerns;

ecological concerns relating to natural resources, systems, communities, freshwater channels and streams and their natural balance, functioning of estuary systems, etc.

Socioeconomic and Cultural Factors

Some aspects to consider in this area include the following:

Demographic aspects: growth rate of population and community, population trends, migration movements

Economic base: employment trends, income distribution aspects, job opportunities, labor movement aspects, welfare system, profile of health care system, profile of cultural aspects and lifestyles

Cultural and religious aspects: historical sites, archaeological aspects, religious connotations

Status of Services

Identify if the proposed development would have a significant impact on, or be constrained by, any of the following characteristics:

- Levels of water and energy supply

- Extent of waste management services

- Transport network services

- Nature of educational services

- Availability of housing and associated services

- Food services

- Shopping and recreational facilities

- Health care facilities and services

- Telecommunication services

- Emergency services

Environmental Pollutions and Hazards

Assess the impact that the proposed development could have in relation to the following factors:

- Air, water, and noise pollution

- Solid or liquid waste and by-product disposal

- Hazards to public

- Workers' safety and risk aspects

- Health and safety aspects

- Various cumulative and synergistic effects

3

DEVELOPING AN EMS MODEL: ISO 14004

In Chapter 2 we outlined the fundamental issues relating to environmental quality management. Also, environmental management practices and principles were briefly described and elucidated to highlight a framework for organizations to manage and control their activities in order to maintain and improve the quality of the environment. With this groundwork we can now embark upon developing a framework for establishing and implementing an Environmental Management System (EMS).

Developing an Effective EMS

For an organization to develop an effective EMS, it needs to complete the steps below. A more detailed outline of these steps is given in Chapter 2, "Preparing an Environmental Profile for EMS."

- Identify the environmental aspect of activities throughout the life-cycle stages.

- Evaluate the environmental impacts of activities.

- Assess the strengths and weaknesses of the current systems to control the environmental loadings of activities.

- Reengineer systems and develop an effective environmental response mechanism.

- Provide appropriate managerial, technical, human, physical, and financial resources to maintain the system.

- Continuously improvise and improve the system.

Environmental Management Principles

In addition to the tangible requirements for developing an effective EMS, there are some fundamental environmental management principles that should become an integral part of an organization's operability. Indeed, without these principles, systems will not be sustainable. Some of these guiding principles are:

- Commitment to do good to the environment

- Minimizing activities that are unfriendly to the environment

- Designing environmentally benign products and services

- Improving product quality and performance

- Working towards sustainable developments

- Using environmentally safe energy resources

- Utilizing natural resources sustainably

- Practicing the 4 R's: reduce, reuse, recycle, recover

EMS MODEL: ISO 14004

We will now develop an environmental management system based on the model given in the international standard *ISO 14004: Environmental Management Systems—General Guidelines on Principles, Systems, and Supporting Techniques.* ISO 14004 is a guidance standard that has been developed to provide generic descriptive guidelines on establishing, maintaining, and improving environmental quality systems. The basic premise of the ISO 14004 guidelines is to provide assistance to organizations for developing an effective EMS to improve environmental quality aspects of their activities. It also provides practical advice on how to effectively initiate, improve, and sustain an environmental quality management system. By following these guiding principles, the organization can lay the foundation for a sound environmental management system and prepare itself for accreditation and compliance with any EMS certification program or regulatory requirement.

The basic EMS model in ISO 14004 is compartmentalized into five implementation phases (see Figure 3-1).

By following these five implementation phases, a suitable EMS model can be developed commensurate with an organization's needs

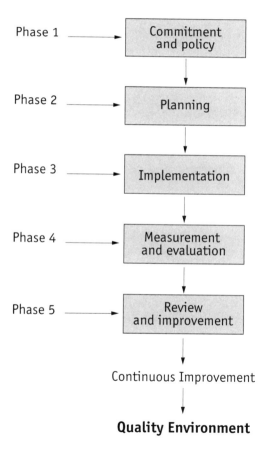

Figure 3-1. ISO 14004: EMS Model

and operability framework. The salient features of the system elements in these phases are outlined below.

Phase 1: Commitment and Policy

General

- Identify the activities that impact the environment.

- Prioritize the activities with respect to the significance of their impact.

- Focus attention (responsibilities, resources) on those activities which have the most significant impact on environment. Then, systematically extend EMS implementation to all activities.

- Integrate all activities relating to quality as well as environmental management into the overall business plan.

Top Management Commitment and Leadership

- Identify top management commitment and develop avenues for the commitment to be known to everyone in the organization.

- Identify and assign leadership roles and responsibilities.

- Encourage participative management.

Initial Environmental Review

- Conduct complete environmental scanning to identify all activities and their environmental impacts and liabilities.

- Evaluate and document significant environmental issues and concerns.

- Evaluate performance of environmental issues against relevant standards, criteria, practices and principles.

- Identify gaps and deficiencies.

- Identify existing practices and procedures to address environmental issues.

- Improve upon the current systems to ensure effective system performance.

- Obtain relevant input and feedback from all concerned parties.

Environmental Policy

- Establish integrated policies, mission, vision, and objectives to address issues relating to both the product quality and environmental issues.

- Communicate these commitments to all levels of the organization.

- Ensure that the policy is understood and implemented at all levels of the organization.

- Ensure everyone's commitment and accountability to the policy.

Phase 2: Planning

General

- Draw up a plan to fulfill environmental policies, commitments, and objectives. The plan should identify environmental aspects and their impact, acceptance criteria, legal requirements, environmental objectives and targets, and the environmental management plans and programs.

Identification of Environmental Aspects and Evaluation of Associated Environmental Impacts

- Identify the environmental aspects and the significant environmental impacts associated with activities, products, and services.

- Identify past, current, and potential impact of activities on the environment. The environmental concerns should be designated with respect to their scale of impact, severity of impact, probability of occurrence, and permanence of impact.

- Environmental aspects, as a minimum, should also include identification of the potential regulatory, legal, and business exposure affecting the organization, as well as impacts on health and safety, and environmental risk assessment.

Legal Requirements

- Establish and maintain procedures to identify and have access to all legal and other requirements associated with the environmental aspects.

- Ensure that environmental requirements of all levels of the regulatory bodies are accommodated.

- Consider regulations pertaining to all aspects of the associated activities, such as: physical characteristics, ecosystem, socioeconomic and cultural aspects, infrastructure services, environmental pollution, risk and hazards, and health and safety aspects.

Internal Performance Criteria

- Develop internal performance criteria to measure system effectiveness against established goals.

- Evaluate performance of all aspects of the system, including management commitment, employee responsibility and accountability, suppliers capability, operational procedures, waste management, emergency response capability, and risk management.

Environmental Objectives and Targets

- Establish environmental objectives commensurate with the environmental policy.

- Establish targets to achieve these objectives.

- Ensure that the targets are measurable and act as environmental performance indicators.

- The objectives and targets should be periodically reviewed and revised to ensure that the desired improvement levels are achieved.

Environmental Management Plans and Programs
- Develop long-term as well as short-term environmental management plans and programs.

- The plans should identify schedules, resources, responsibilities, objectives, and targets.

- The environmental plans can be integrated into other quality and/or business plans.

- The plans should be dynamic in nature and should be revised regularly to reflect changing priorities.

Phase 3: Implementation

General
- To achieve effective implementation of the EMS, the organization should develop the necessary capabilities and support mechanisms, and focus its attention on aligning its people, systems, strategies, resources, and infrastructure.

Ensuring Capability
- **Resources—Human, Physical, and Financial:** Allocate appropriate human, physical, and financial resources to achieve stated objectives and targets.

- **EMS Alignment and Integration:** Systems are generally more sustainable when they are interactive and integrated in nature. The EMS should be integrated with other business plans or quality systems.

* **Accountability and Responsibility:** Appoint a senior person with designated responsibility and authority for the integrated quality-environmental system. Appropriate responsibilities must be defined at all levels in the organization. Employees should be accountable for the performance of the EMS within the scope of their responsibility.

* **Environmental Values and Motivation:** Senior management should play a key role in communicating a shared vision and common set of values. Participative management, people empowerment, and recognition would result in motivated work force.

* **Knowledge, Skills, and Training:** Procedures should be established to identify skills required to achieve environmental objectives and targets. Training and development opportunities should be available to all employees. Employees should be selected on the basis of their knowledge, training, and experience.

Support Action

* **Communication and Reporting:** Establish appropriate means to communicate and report, internally and externally, on the performance of activities and procedures for the effective functioning of the EMS. Effective communication raises the awareness level of the employees as well as serves to motivate and encourage open and participative involvement.

* **EMS Documentation:** Operational processes and procedures must be appropriately documented and maintained at all levels of the organization. The documents should be readily identifiable and revised as appropriate.

- **EMS Records and Information Management:** All pertinent data, documentation, and information must be effectively managed. Environmental quality records must be properly maintained and readily available for performance evaluation. Some of the important records pertain to aspects such as: legislative and regulatory requirements, environmental aspects and impacts, training activity, inspection and calibration activity, monitoring data, records of nonconformance and corrective action, audit results, management reviews of the EMS.

- **Operational Controls:** Establish and maintain operational procedures and controls to ensure that the level of environmental performance is consistent with policies, objectives, and targets.

- **Emergency Preparedness and Response:** Establish emergency response capability, plans, and procedures to ensure appropriate and timely response to any environmental incidents and emergency situations.

Phase 4: Measurement and Evaluation

General

- Continuously measure, monitor, and evaluate key activities of the EMS.

Measuring and Monitoring (Ongoing Performance)

- Establish a system for measuring and monitoring actual performance against environmental objectives and targets.

- Analyze the results to identify gaps and deficiencies.

- Take appropriate corrective/preventive actions.

- Establish environmental performance indicators that are objective, verifiable, and reproducible.

Audits of the Environmental Measurement System

- Establish audit procedures and conduct regular audits of the EMS to ensure conformance to planned arrangements and activities.

- Audits, whether conducted by the internal staff or by an external auditor, must be objective and impartial.

- Audit reports should be submitted as per the audit plan.

Phase 5: Review and Improvement

General

- Continuous improvement processes should be applied to the EMS to achieve overall improvement in environmental performance.

Review of the Environmental Management System

- Conduct management reviews of the EMS at appropriate intervals to ensure its continuing suitability and effectiveness.

- Management reviews should include: review of objectives and targets, audit reviews, system effectiveness, compliance with regulatory requirements, training activities, nonconformance and corrective actions, etc.

Corrective and Preventive Action

- Results of nonconformities and audit findings should be recorded and reported to the responsibility centers.

- Appropriate corrective and preventive action should be taken and its implementation ensured through follow-up action.

Continual Improvement

- Establish a suitable process for continual improvement of the EMS.

EMS IMPLEMENTATION: A ROAD MAP

The success of the system depends on it being commensurate with the organization's existing systems and current procedures and practices. The system must be simple and user-friendly and not be imposed on people. A serious effort should be made to make it a people-developed and people-empowered system. In maintaining a successful system, everybody should be collectively involved and responsible.

Organizations can customize their environmental system by using the four step P-U-R-I: Plan-Upgrade-Record-Improve Road Map, as schematically presented in Figure 3-2. This simple working road map complements the main features of the EMS model of ISO 14004.

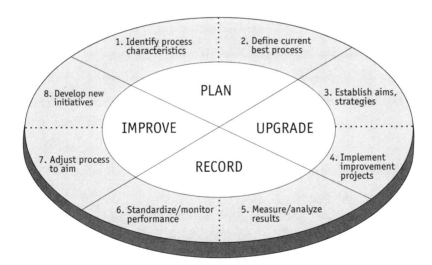

Figure 3-2. "Puri" Process Enhancement Wheel

Step 1: PLAN

- Develop an environmental policy, highlighting the organization's vision, mission, core values, beliefs, and guiding principles.

- Integrate environmental policy with other policy statements or business statements of the organization.

- Communicate commitment to policy to all employees and ensure that the policy is clearly understood, implemented, and maintained at all levels of the organization.

- Arrange for a management orientation and awareness session on environmental aspects of quality.

- Appoint an EMS coordinator, with defined responsibility and authority, to coordinate and manage all aspects of the environmental quality system.

- Establish a steering committee to oversee the overall functioning and improvement of the EMS.

- Establish a mechanism to obtain feedback from the stakeholders and ensure that their needs and demands are accommodated in the environmental policy, objectives, and targets.

- Identify and ensure compliance with the legal and regulatory requirements concerning the impact of activities on the environment, purchased materials, products, and services.

Step 2: UPGRADE

- Develop a master list of activities of the organization and identify the environmental aspects of these activities throughout the various life-cycle stages.

- Develop a profile of environmental impacts of activities.

- Evaluate the suitability of current procedures and practices to control the environmental impacts of activities, for example, emissions to air, releases of water, contamination, and waste management.

- Conduct a gap analysis to evaluate strengths and weaknesses of the current system in responding to environmental issues and problems.

- Develop a master EMS implementation plan with time schedules, responsibilities, control points, review procedures, etc.

- Allocate appropriate resources: human, financial, and technological.

- Develop appropriate documentation and procedures.

- Provide appropriate training and education at all levels of the organization.

Step 3: RECORD

- Maintain documented procedures at requisite locations and ensure that they are continuously revised and updated.

- Establish critical control points in all activity areas to control the environmental effects.

- Establish emergency response procedures.

- Develop a suitable document control system.

- Establish a system for maintaining environmental quality records.

Step 4: IMPROVE

- Establish process improvement teams.

- Conduct process capability studies.

- Measure, monitor, and analyze key environmental characteristics at all critical control points of processes.

- Establish and maintain a suitable program for audit of the EMS system.

- Provide continuous training throughout the organization on the control, maintenance, and improvement of environmental quality.

- Establish appropriate automated/computerized systems or technologies to monitor and control the processes.

- Conduct performance evaluation of the EMS at regular intervals.

- Implement corrective and preventive action mechanisms to improve the system.

- Review and revise standard operating procedure vis-à-vis the corrective/preventive action.

- Conduct management review of the system at regular intervals.

- Establish a tangibly identifiable program of continual improvement of the EMS.

- Continuously partner with suppliers, subcontractors, and stakeholders to identify environmental concerns and take appropriate measures to improve the system.

- Identify improvement opportunities, establish strategic initiatives, and allocate requisite resources.

- Focus on never-ending cycle of continuous improvement.

Benefits of a Proactive EMS

A revolution for the protection of the environment is on the horizon. The activities of the next century will be overwhelmingly linked to the paramount issues of environmental management and human health and safety, issues that are intricately interwoven with our ecosystem. Consequently, environmental quality is an issue that commands everyone's attention. Organizations who undertake activities, produce products, or deliver services that have an impact on the environment must clearly understand their roles and responsibilities in ensuring environmental quality and safety. They have a responsibility toward society—the society in which they themselves exist.

Notwithstanding the societal obligation, organizations realize many tangible and intangible benefits from effective environmental management systems. Companies striving for quality and excellence are cognizant of these economic benefits and they will always operate in a proactive rather than reactive mode in addressing these important issues. As illustrated in Figure 3-3, a proactive EMS will enhance a company's ability to achieve significant competitive advantage through operational and production efficiencies, as well as increase acceptability of products and services by the public. It also will assist with cost control and the minimization of losses. By linking an environmental objective and targets with specific financial outcomes, an optimal utilization of natural resources will occur. And by meeting the customers' and stakeholders' environmental and financial expectations, the company will enhance its market credibility and market share. With a proactive EMS, a company will also have an effective waste management system already in place; a reduction in liabilities; improved industry and government relations; new opportunities for improved technology; and enhanced service to society by contributing to a better quality environment.

ENVIRONMENTAL PERFORMANCE EVALUATION (EPE)

Environmental performance evaluation (EPE) is an integral part of any environmental management system (EMS). It is a tool that can be effectively used to evaluate the performance of the EMS for continuous improvement of the system. An effective EPE process can help an organization objectively verify if the systems are operating in concert with the stipulated environmental policy, objectives, and targets. Once an organization has successfully implemented an environmental

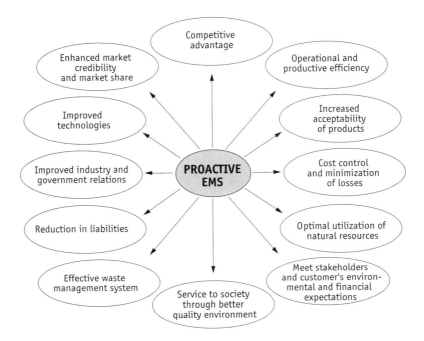

Figure 3-3. Benefits of a Proactive EMS

management system, it should develop an appropriate EPE procedure as part of the management review activities or integrate the EPE process into the measurement and evaluation phase of the EMS.

Environmental performance evaluation is distinctly different from other assessment tools, such as reviews and audits. Whereas environmental management system reviews and audits are designed to provide information about a functioning EMS at a specific point in time, the purpose of environmental performance evaluation is to collect and analyze ongoing data and information to provide a current, up-to-date, and continuous evaluation of the environmental system as well as showing trends over time. An EPE helps to identify system deficiencies and environmental performance indicators on the basis of which the decision-makers can implement corrective actions and continual improvement initiatives.

The international standard, *ISO 14031: Environmental Performance Evaluation* defines EPE as follows:

> **Environmental Performance Evaluation (EPE):** a process to measure, analyze, assess and describe an organization's environmental performance against agreed criteria as appropriate for management purposes.

As an internal management tool, the EPE process is implemented within an organization's environmental management system to assess and describe the organization's environmental performance related to its environmental policy, objectives, and targets, and to assist in achieving continual improvement of the EMS.

Areas of Application for EPE

Air Emissions: Release of gases, particulates, or toxic or odorous substances to the air that may cause depletion of stratospheric ozone or create ground level ozone which may contribute to smog formation, and releases of pollutants or other fugitive emissions or corrosive and acidic substances that may adversely affect the health and safety of human beings, the surrounding environment, or the natural ecosystem.

Solid Wastes: Solid wastes produced at all stages of a product's life-cycle: solid or liquid products or materials deposited in landfills whether before or after incineration, composting, recovery, or recycling; extraction of raw materials and other resultant wastes or unwanted side products arising out of manufacturing processes; potential waste emanating from recycling and reversing processes; transportation, storage, distribution, and disposal of products.

Water Discharges: Discharges of pollutants from point or diffuse sources to sewers or natural waterways which may give rise to adverse environmental effects, including toxic, radioactive, oxygen depleting, or tainting substances; after water treatment devices or accidental releases of effluents to a watercourse; either surface or ground water discharges of nutrients that may cause eutrophication of natural water; discharges to waterways that may cause various pollutant effects on aquatic ecosystem.

Other Environmental Emissions/Releases: Radiation, waste heat, odor, noise and vibration, and other releases typical of manufacturing, transportation, and waste disposal processes.

Interactive Steps in the EPE Process

Outlined below are the basic interactive steps in the EPE process.

1. Selection of the overall evaluation categories within each of the life-cycle stages of activities

2. Development of environmental performance indicators

3. Measurement of results

4. Analysis of data and information

5. Assessment of results

6. Reporting of results to management for review and action

7. Implementation of corrective action

8. Identification of opportunities and initiative for continuous improvement of the EMS

Advantages of an Effective EPE

The major advantage in implementing an effective EPE process is that it can generate information which may be utilized to:

- Better understand the nature and significance of the environmental aspects and impacts of activities

- Obtain accurate information on the magnitude and intensity of environmental risks associated with activities, products, and services

- Optimize the allocation of managerial, technical, and financial resources

- Identify nonconformities and their root causes, and implement corrective and preventive actions

- Measure, analyze, evaluate, and highlight the organization's environmental performance against its policy, objectives, targets, and improvement programs

- Ensure consistency among various units in the organization with regards to implementation of policies, objectives, targets, and procedures

- Ensure compliance with regulatory and legislative requirements

- Provide an objective basis for appropriate reporting and communication to internal and external interested parties and stakeholders

- Generate information which can be used to assess the organization's overall performance and to facilitate the improvement efforts

- Identify potential business opportunities

- Help maintain continual improvement focus

EPE Design and Functional Capability

The success of an EPE process is dependent on its design and functional capability. The process must be developed with great care and tenacity to ensure that it will fulfill its intended purpose. A well-designed EPE process must:

- Adequately address the organization's needs and requirements

- Be appropriate for the management purposes for which it is designed

- Support the organization's environmental policies, objectives and targets

- Generate objective evidence on the organization's environmental performance

A well-developed EPE process should also:

- Have the capability to identify compliance with regulatory requirements

- Generate information that can be appropriately communicated to the stakeholders

- Be compatible with other management information systems

- Provide understandable, relevant, objective, and verifiable qualitative and/or quantitative information appropriate to all aspects of environmental burdens

- Provide adequate information for all life-cycle stages of activities.

ELEMENTS OF AN EPE PROCESS: ISO 14031

In this section, we will present the basic framework of an EPE process as outlined in the international standard, ISO 14031. Guidelines are also provided for these elements which can be productively used to establish and implement an effective environmental performance evaluation system.

5.1 Planning for EPE

Plan the EPE process during the phase when the environmental policy, objectives and targets are being established. The EPE process should ensure that:

- The environmental policy, objectives and targets are appropriately expressed into operational activities that can be adequately verified and assessed.

- Appropriate evaluation areas are clearly identified.

- Relevant environmental performance indicators are suitably identified to measure environmental performance.

5.2 Selection of Evaluation Areas

There are three basic areas where the EPE process should be utilized:

1. Management System

2. Operational System

3. State of the Environment

The extent to which these areas would be evaluated depends on the needs and priorities of the organization. Also, the EPE process should be so comprehensive as to collectively address and build upon linkages across the three evaluation areas.

5.2.1 Management System

- Activities for evaluation within the management system area include: procedures, processes, planning, resource allocation, operational controls, feedback and verification of performance information and results, compliance with legal and stakeholder requirements. In an organization with several levels of functional hierarchy, the EPE process must be designed to accommodate the needs of all levels, either individually or collectively.

- Since a management system interacts with and is linked to the operational system, the EPE process, when applied to management system areas, should take into consideration all requisite interactive elements of the operational aspects.

- The EPE process should yield environmental performance indicators describing the strengths and weaknesses of the EMS and these should provide the basis for improving the EMS.

5.2.2 Operational System

- Activities in this category include: design and operation of plants, equipment mass and energy flows required for the generation of products and services, discharges and emission of activities, and all the other interactive and interlinked activities of plants and sites.

- Environmental performance indicators should identify and evaluate the organization's progress in meeting its environmental policy, objectives, and targets. This information should be used to make improvements to the functioning of the operational system.

5.2.3 State of the Environment

- Activities in this area involve functions beyond the boundaries of the operating facility, such as distribution and waste systems, changes in the overall environment, health of people and animals, entities of the ecosystem, and ozone depletion levels.

- The organization should cooperate and work together with other agencies who carry out capability studies on the state of the environment. The collective information and results should be used to make improvements either in the internal environmental management system and/or in the external activities associated with the environment.

5.3 Selection of Environmental Performance Indicators (EPIs)

- The environmental performance indicators (EPIs) that are to be used in the environmental performance evaluation (EPE) process must be identified at the same time as an EMS and its objectives and targets are established.

- Ensure that the EPIs are relevant to the organization's policy, objectives, and targets.

- Ensure that the views of stakeholders and interested parties are accommodated in setting of priorities, environmental policy, objectives, and targets, and in the selection of appropriate and relevant EPIs.

5.3.1 Factors to Consider in Selecting EPIs
- Since EPIs are necessary and instrumental in generating concise, objective, and reliable environmental performance data, their identification and selection must be done with great care and tenacity. One approach to ensure that the EPIs are reliable and relevant is to assess their performance with respect to the following three categories:

  – data reliability

  – EPI relevance

  – usefulness to decision-makers

- To achieve a balanced assessment of the evaluation areas, the EPI's must reflect the needs, concerns, and expectations of the internal and external interested parties and stakeholders, as well as the organization's environmental policy, objectives, and targets. The following matters are relevant to environmental performance and may be used in the selection of appropriate EPIs:

– environmental objectives of the organization

– legal and regulatory requirements

– environmental aspects of activities and their associated impacts and risks

– the operating framework of manufacturing activities

– needs and expectations of interested parties and stakeholders

– technological and financial aspects

– business and environmental strategy and potential growth

5.3.2 Types of Environmental Performance Indicators

Management System Indicators

• Customer satisfaction indicators

• Financial indicators

• Reliability of data and information

• Effectiveness of EMS implementation

• Rate of nonconformances

• Effectiveness of corrective action implementation

Operational System Indicators

• Energy consumption

• Resources expended

- Efficiency of material and energy utilization

- Emission and waste indicators

State of the Environment Indicators

- Indicators related to environmental impacts and risks

- Emergency response preparedness

5.4 Data Collection and Measurement

- Reliable data collection and measurement systems are essential to the effective functioning of an EPE process. The following types of data may be generated:

  – absolute data

  – relative in compound data

  – aggregated data for related factors

  – indexing

  – qualitative data

- Data must be properly collected. For each performance category, at least one EPI must be selected and applied in order to measure environmental performance in that area.

5.5 Analysis of Data

- Recorded data should be collected, aggregated, manipulated, and analyzed in various ways in order to evaluate performance of the EMS against predefined targets.

- Analysis may also include relative measures, such as assigning rankings and weightings to information. This information may be stated in qualitative or quantitative terms.

5.6 Evaluation and Assessment

- Results are to be assessed to identify whether the organization has achieved its environmental performance targets. Also, evaluate the costs incurred in meeting environmental objectives.

- The aspects that need to be considered in the evaluation and assessment process include:

  - organization's environmental policy, objectives, targets, and initiatives

  - information with respect to environmental performance in selected evaluation areas

  - information regarding the severity and extent of environmental risks and impacts

5.6.1 Use of Assessment Information
The information generated through the evaluation and assessment process is used as a basis for:
- Development of recommendations on performance improvements

- Changes necessary to improve the EPE process

- Future action on areas where deficiencies have been identified

5.7 Communications and Reporting

Results of the assessment and evaluation should be reported to the decision-makers, employees, and external stakeholders. These communications should:

- Demonstrate management commitment to environmental quality and safety

- Adequately deal with environmental concerns of activities, products, and services

- Raise awareness about policies, objectives, targets, and initiatives

- Provide appropriate and reliable information about the organization's environmental management system

- Provide information on environmental performance of the EMS

- Provide management with objective and accurate information on the performance of the EMS for review and action

6.0 Continual Improvement of the EPE Process

6.1 Redefining the Use of Indicators

The EPE process should be designed to provide suitable recommendations on how to:

- Develop effective environmental performance indicators

- Achieve improvements in data quality, integrity, and availability

- Make appropriate changes to analysis and evaluation process

6.2 Identify Business Risks and Opportunities

The EPE process should be capable of identifying business risks and opportunities for management to evaluate as they develop effective strategies for improvement and growth.

ENVIRONMENTAL RISK ASSESSMENT (ERA)

In the preceding sections, we outlined the guidelines for developing and implementing an environmental management system (EMS) and carrying out an environmental performance evaluation (EPE) of the system. We shall now provide the basic procedural framework of another important tool for the improvement and monitoring of the EMS—the Environmental Risk Assessment (ERA) process.

As we know, during the development and implementation of an EMS, we have to identify the environmental aspects of activities during the various life-cycle stages and evaluate their impacts on the environment. In doing so, we have to consider all possible entities that may be instrumental in causing these environmental effects. For example, the environmental effects may be due to some factors relating to environmental policy, governmental regulations, emission standards, stakeholders expectations, or biota of the region. In determining the likelihood and seriousness of implications resulting from these situations and factors, it is essential to make a number of judgments. These judgments can, indeed, create disagreements and conflicts between businesses, regulators, and stakeholders on the viability and seriousness of the environmental impacts. This is where the process of environmental risk assessment plays a significant role. The ERA process helps to make rational, objective, and defensible decisions regarding the seriousness of the environmental effects, and minimizes any disagreements arising out of conflicting opinions.

Reasons for an ERA

Normally, the environmental risk assessment process is carried out for one or several of the following reasons:

- As part of the management commitment for enhancing, sustaining, and protecting the environment

- Whenever a significant change occurs in the business activity, e.g., change in material structure and composition, energy flows, or changes in the flora or fauna of the ecosystem

- If required by changes in the regulatory requirements

- If initiated by some specific requirements of the trade obligations

- If mandated by contract requirements of the customer

Uses of ERA Process

Environmental risk assessment is a process that can be used to:

- Identify real and potential environmental effects of activities, products, and services

- Evaluate the magnitude and seriousness of these effects

- Identify the probability of their occurrence

- Highlight the extent of consequences resulting from these environmental impacts

ERA Results

The results of an ERA process can be used to do the following:

- Compare the environmental effects and their consequences to the stipulated regulatory or stakeholders's requirements for compatibility and any appropriate action.

- Take timely action to minimize either the likelihood of occurrence of the environmental impacts or their resulting consequences.

- Utilize the information for developing strategic plans and directions for improving the EMS.

ENVIRONMENTAL RISK ASSESSMENT PROCESS

The main objective of an ERA study is to analyze the environmental risks associated with activities; evaluate the significance of their impact; compare the potential impacts with the requisite standard requirements; calculate the likelihood of occurrence for these impacts; and, implement measures to control the environmental effects. The environmental risk assessment process can, therefore, be compartmentalized and discussed under four stages (see Figure 3-4).

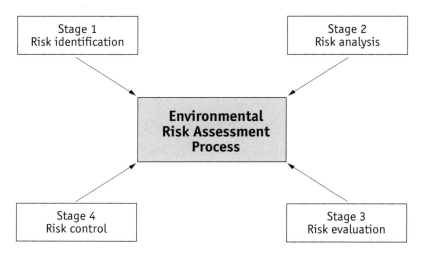

Figure 3-4. Environmental Risk Assessment Process

Requirements for an ERA Study

As is the case with any assessment project, a study plan must be established for executing and completing the project. Some of the basic requirements for the ERA study include the following:

- Establish a documented set of objectives and goals for the study. These goals should be compatible with the nature and type of activity, requirements of the stakeholders, and management's business plan. The assessment goals should also be responsive to:

  – the needs of those who will use the results of the assessment

  – the nature and type of existing as well as potential risks

– implications regarding subsequent ERA studies and monitoring of activities

- Establish and document the details of methodology to be used for conducting the study. This includes:

  – developing a format for the questionnaire

  – identifying data collection and analysis methods

  – specifying technical parameters of the study, including volume, nature, and quality of data required; selection of model and establishment of other requisite assumptions; methods of handling subjective and objective data; definitions of acceptable levels of risk and uncertainty; and any weighting criteria to be used in analyzing the risks

- The study should adequately address:

  – a clear understanding of the nature of hazard and the environmental system at risk

  – technical capability to provide relevant measurements and information to alert to changes in the environment

  – core values of stakeholders and society regarding the important aspects of the environment

  – comparability between risks and environmental endpoints and standards.

- The risk assessment team members should:

  – be competent in the conduct of the study

  – be proficient in the evaluation procedures

  – be qualified in the knowledge and understanding of risks

  – have a thorough knowledge of the subject under investigation

- Establish an appropriate documentation system for all aspects of the study project. Documentation should include:

  - study methods, study assumptions and the results of each step of the study

  - data and information gathered and its analysis

- Documentation should also include any uncertainties in either the analysis or the evaluation in terms of: any inherent variability and uncertainty regarding the parameter values; any gaps in technical and scientific knowledge and theory; any other relevant information that may bear an impact on the validity of findings

- Establish a format for the reporting of the findings and the verification of results

- Establish procedures for corrective action and follow-up activities

The procedural details for the four stages of ERA study are outlined in Table 3-1.

Table 3-1. Procedural Details for the Four Stages of ERA

Stage 1: Risk Identification

- Describe a general profile of the environment under study.

- Describe the activities, products, and services produced in the environment.

- Identify all the physical, biological, chemical, and energy inputs and outputs of activities, including their nature, volume, and any other characteristics that may bear an influence on the environment.

- Identify the environmental aspects of activities.

- Describe the processes used within each activity that may have an influence on the environment.

- Identify the potential hazards and their environmental impact during all stages of the life-cycle of activities.

Stage 2: Risk Analysis

- Analyze each hazard and outline the cause, pathway, and mode of operation of the hazards; and the potential negative environmental consequences of the hazards.

- Estimate the effects of hazards, identifying the components of the environment that might be exposed to the hazards from accidental or process release of the chemicals, and the extent of effects of these hazards.

Stage 3: Risk Evaluation

- Analyze the data and information relating to the hazards.

- Calculate probabilities of occurrence of the hazards.

(cont'd)

Compare the results of the assessment with endpoints. Endpoints, with respect to environmental risks, are generally developed by management in consultation with regulatory standards and needs of the stakeholders. Endpoints are normally expressed as regulated or non-regulated standards of general or particular application, and are based upon:

– technical and scientific information and opinion

– regulatory policies and priorities

– societal values and expectations

Examples of endpoints for environmental risk are: emission standards, water quality standards, waste disposal standards, or any given levels of biodiversity.

- Determine the acceptability of the risks in relation to the endpoints. Other factors to be considered for evaluating the risks and determining their acceptability include: economic value of the environmental components at risk; distribution of the risks, costs and benefits associated with the activities being assessed; implications to the stakeholders; and options available to those exposed to the risks.

Stage 4: Risk Control

- If the assessment findings meet the endpoint requirements, the study can be stopped and the organization can resume their activities, with continuous monitoring of the system. The decision to stop the ERA process will be based on the following:

– level of confidence in the results

– assurance that the study methods were adequate and error-free

– sensitivity of results to changes in the assessment procedures or changes in the activities being assessed

– need for further studies and the availability of resources required to carry out such studies

- If the endpoints are not being met, or it is felt that further ecological and economic benefits might still be obtained from modifications to the activity, then either the ERA study should be continued or a risk management study should be initiated.

- Exercise continuous monitoring and control of activities.

- Continuous EMS reviews should be conducted to improve the system's effectiveness.

4

OTHER EMS STANDARDS/PROTOCOLS

Environmental concerns have been a global issue over the past few years. These concerns gave rise to the issuance of many guiding principles for environmental management as well as business charters for sustainable development. In this chapter, we will cover in some detail the framework and principles associated with the British Standard, BS 7750, as well as the Canadian Standard, CSA-Z750-94. We will also present an outline of the system elements of an Environmental Management System (EMS) as discussed in the South African Standard, SABS-0251. (An outline of twelve international protocols is given in Appendix A.) This information will provide a useful source of reference and will highlight the commonality of system requirements inherent in all standards and protocols associated with environmental quality management. It will also serve as an additional source of guidance for establishing your environmental management system.

BRITISH STANDARD: BS 7750

The *British Standard, BS 7750: Specification for Environmental Management Systems*, is perhaps the oldest document on the subject of environmental quality management. It specifies core requirements for an EMS, as well as provides guidance on the implementation and assessment of the system elements. As a prescriptive standard, it supports EMS certification schemes and as such, is used for certification purposes in many countries. Organizations who are already utilizing the guidelines and specifications given in BS 7750, whether for establishing an EMS or for certification purposes, will find the following narrative and explanation of system elements particularly useful.

We will now elucidate the salient features of the system elements of the standard which can be utilized to develop an EMS model. The reader is, however, advised to consult the standard for further details.

BS 7750: Specification for Environmental Management Systems—Elements

4.1 Environmental Management System

- Establish and maintain an EMS, commensurate with environmental policy, objectives, and targets to ensure that the environmental burdens of the organization's activities are under control. The EMS should be supported by appropriate documented system procedures and their effective implementation.

- The implemented EMS should also be in compliance with any regulatory requirements or codes of practice.

4.2 Environmental Policy

- Establish and document the environmental policy.

- The environmental policy can be integrated with any other broader corporate policy or quality policy.

- The environmental policy should:

  – adequately accommodate and address the environmental aspects and impacts of the activities

  – identify management commitment to environmental protection

  – be communicated to all levels of the organization

– identify for the stakeholders how the policy would be translated into objectives and measurable targets

4.3 Organization and Personnel

4.3.1 Responsibility, Authority and Resources

- Define and document appropriate responsibilities and authority for the effective functioning of the EMS.

- Ensure that the personnel who are responsible for managing, performing, and verifying activities which have a significant impact on the environment have the freedom and authority to:

  – access adequate resources—human, financial, technological—to ensure effective functioning of the system

  – initiate action and recommend solutions for any problems or deficiencies that may occur in the implementation of the system

  – verify that the system is operating efficiently

  – initiate action in an environmental emergency situation

4.3.2 Verification Resources and Personnel

- Establish and maintain procedures for the verification of activities.

- Provide adequate resources and personnel for verification of system effectiveness.

4.3.3 Management Representative

- Appoint a management representative, with defined responsibility and authority, to oversee and ensure that the EMS is effectively implemented and maintained.

4.3.4 Personnel, Communication, and Training

- Establish effective communication channels to ensure that employees at levels of the organization are completely cognizant of the importance of the EMS in terms of:

  – compliance with the environmental policy, objectives and targets

  – significance of the environmental burdens of their activities

  – their roles and responsibilities in achieving the stated environmental objectives

- Establish effective training and development programs to ensure that personnel whose work may have a significant impact on the environment are adequately trained.

4.3.5 Contractors

- Establish procedures to ensure that the suppliers and subcontractors clearly understand the environmental management system requirements of the organization, and are aware of the importance of compliance with these requirements.

4.4 Environmental Effects

4.4.1 Communications

- Develop a mechanism to ensure that all pertinent information regarding the environmental issues is effectively disseminated to and from the relevant interested parties.

4.4.2 Environmental Effects Evaluation and Register

- Establish procedures to identify, evaluate, and record the environmental aspects of activities. Consideration should be given to activities such as: emissions to the air; discharge to water; solid and other wastes; contamination; use of natural resources such as land, water, fuels, and energy; other environmental loadings, including burdens on the ecosystem.

- Categorize and rate the environmental aspects in terms of the magnitude of their impact, i.e., having significant impact or having no impact.

- Identify and document environmental impacts under all possible types of operating conditions.

4.4.3 Register of Legislative, Regulatory and Other Policy Requirements

- Establish a register of all legislative, regulatory, and other codes of practice applicable to environmental protection with regards to the organization's activities, products, and services.

4.5 Environmental Objectives and Targets

- Establish environmental objectives and targets.

- The objective must be consistent with the environmental policy, needs and requirements of the stakeholders, and all relevant legislative and regulatory requirements.

- The objectives and targets should be measurable to the extent possible and identify commitment to continual improvement.

4.6 Environmental Management Program

- Establish and maintain an environmental quality management program to ensure sustainable levels of environmental protection.

- Establish separate programs to address new product development as well as existing activities.

- Any EMS program should clearly identify: the environmental objectives to be achieved, the means of achievement, the extent of achievements made, and continuity of efforts for sustainable developments.

4.7 Environmental Management Manual and Documentation

4.7.1 Manual
- Establish and maintain an EMS manual, in paper or electronic form. The manual should address, as a minimum, the following activities:

  - environmental policy, objectives, targets, and programs

  - appropriate responsibilities and authority at all levels of the organization

  - framework and functionality of standard operating procedures

  - cross-functional interface of entities, programs, systems, and procedures within the organization

  - environmental emergency response procedures

  - mechanism for controlling, monitoring, and evaluating the system effectiveness

  - commitment to continuous improvement

4.7.2 Documentation
- Establish and maintain standard operating procedures and all requisite documents for the effective functioning of the EMS.

- Establish a proper document control system, identifying responsibilities and authority for maintaining, reviewing, and deleting the procedures.

- All documents must be properly identified and maintained and readily retrievable.

4.8 Operational Control

4.8.1 General

- Establish defined responsibility and authority to ensure that control, verification, measurement, and testing activities are properly coordinated and effectively performed.

4.8.2 Control

- Establish procedures to ensure that all activities that have an impact on the environment are performed under controlled conditions. Control conditions encompass:

  – documented procedures, standard operating procedures, work instructions

  – monitoring and control of relevant process parameters and product characteristics

  – maintenance and approval of processes and equipment

  – criteria of workmanship and acceptance stipulated in written procedures, practices and standards

4.8.3 Verification, Measurement, and Testing

- Establish and maintain procedures for verification of activities to ensure compliance with specified operational as well as regulatory requirements.

- Establish calibration procedures for the control of inspection, measuring, and test equipment.

- Maintain adequate records of verification activities and calibration status.

4.8.4 *Noncompliance and Corrective Action*

- Establish procedures for the identification of nonconformances and taking of corrective and preventive action.

- Define and document appropriate responsibilities and authority.

- Maintain records of corrective action and record any changes in procedures resulting from corrective and preventive action.

4.9 Environmental Management Records

- Establish procedures for the identification, collection, indexing, filing, storage, maintenance, and disposition of EMS records.

- Records should be legible and identifiable to pertinent activity, product, process or service.

- Records should be safely stored and readily retrievable.

- Records should be analyzed to evaluate system performance.

4.10 Environmental Management Audits

4.10.1 *General*

- Establish and maintain procedures for regular audits of the EMS system.

- The audits should identify conformance of activities to documented procedures and the effectiveness of the EMS in fulfilling the organization's environmental mandate.

4.10.2 Audit Program

- Establish a formal audit program to ensure that all specific activities and areas are being properly audited.

- The audit program should specify frequency of audits, audit schedules, and appropriate responsibilities for audit.

4.10.3 Audit Protocols and Procedures

The protocol for audit requires the following:

- Documentation of audit findings, audit reports, and records.

- Reports on the performance of the EMS.

- Audit personnel must: be independent of the activities audited, have access to appropriate resources for conducting the audit, and be adequately trained in audit procedures.

- Establishment of audit methodology and checklist commensurate with the nature of the function being audited.

- Procedures for reporting audit findings to auditees and ensuring that timely corrective action has been taken.

- Reporting of audit findings to the management for review and improvement of the system.

4.11 Environmental Management Reviews

- Management should review the EMS, at defined intervals, to ensure its continuing suitability and effectiveness.

- Records of system reviews should be utilized to make necessary changes in policy, strategic directions, procedures, processes, and continual improvement programs.

CANADIAN STANDARD: CSA-Z750-94

The *Canadian Standard, CSA-Z750-94: A Voluntary Environmental Management System*, provides guidance on the development and implementation of environmental management systems. It outlines the key elements of an EMS and provides practical advice on implementing or enhancing such a system.

The EMS model and its system elements have been compartmentalized into the following four categories:

Stage 1: Purpose

Stage 2: Commitment

Stage 3: Capability

Stage 4: Learning

The model provides practical guidelines on implementing a sustainable quality system. The salient features of the model are outlined below.

CSA-Z750-94: Environmental Management System—Principles

Stage 1: Purpose (Define Purpose and Establish Plan)

An organization is a sum total of people, system, infrastructure, and resources. It is imperative to define a purpose and a clear sense of direction. This is done through the establishment of environmental policy, objectives, and targets.

Environment Policy: Establish an appropriate environmental policy that would provide the basis for the environmental management system. The environmental policy should be built on the organization's mission, vision, core values, beliefs, guiding principles, and practices. The policy should be commensurate with the needs and requirements of the stakeholders and any requisite legislative or regulatory mandate.

Risk Assessment: Identify the environmental aspects of activities, products, processes, and services. Evaluate the risks and the environmental impacts associated with the activities. Prioritize the objectives and targets vis-à-vis the risks.

Environmental Objectives and Targets: Establish environmental objectives in line with the commitment identified in the environmental policy. Develop a detailed set of measurable targets to achieve the stated objectives.

Stage 2: Commitment (Establish Commitment)

Commitment of the organization and the people who make decisions and take actions for the achievement of environmental quality goals is imperative to the success of an EMS.

Environmental Values: Establish a common set of values and a shared vision towards which all members of the organization can work collectively. The vision should be commensurate with the environmental policy, objectives and goals. All members of the organization should understand the importance of achieving the environmental objectives for which they are responsible and accountable.

Alignment and Integration: Establish an interactive operational framework in which all members of the organization share responsibility and support each other.

Accountability and Responsibility: Appoint a senior person in the organization, with designated responsibility, authority, and accountability, to oversee the effective functioning of the EMS. This person should be responsible for delineating of the appropriate responsibility and accountability to the various functions, monitoring the system effectiveness, and reporting to management on the performance of the system.

Stage 3: Capability Processes (Ensure Capability)

The organization should ensure that requisite resources are continuously provided commensurate with changing and evolving environmental needs and requirements.

Resources—Human, Physical, and Financial: Establish a mechanism for the identification of human, physical, and financial resources necessary for the effective functioning of the EMS and ensure that these are made available to the responsibility centers as appropriate.

Knowledge, Skills, and Training: Establish procedures for identifying training needs and for providing the appropriate training commensurate with the requirements of the job. Employees should have an appropriate knowledge base to perform their functions in an efficient and competent manner. Persons assigned to activities that have a significant impact on the environment should be selected on the basis of their knowledge, experience, and training. Training records should be maintained.

Information Management and Procedures: Establish requisite hierarchy of system documentation to control the functioning of the EMS. Pertinent documents include: Environmental Quality Manual, Environmental Quality System Procedures, and Work Instructions. These documents can be integrated with other quality system documentation. Establish an appropriate management information system for the identification, collection, indexing, filing, storage, maintenance, retrieval, and disposition of pertinent EMS documentation and records.

Stage 4: Learning Processes (Evaluate, Learn, and Improve)

For sustainable developments, it is imperative that the organization is dynamic in nature, capable of effective change management, and focused on continuous improvement.

Measuring and Monitoring: Establish a system for ensuring and monitoring system performance against the environmental objectives and goals. Analyze all pertinent information to determine the degree of success in achieving the desired targets and to take any appropriate corrective and preventive action to enhance system performance.

Communication and Reporting: Establish an effective communication system to report the results of system evaluation and its degree of performance to all concerned parties within the organization and to external stakeholders.

System Audits and Management Review: Establish and maintain suitable environmental system audit programs to confirm the adequacy of the EMS in achieving desired environmental objectives. Management should conduct EMS reviews, at appropriate intervals, to ensure its continuing suitability and effectiveness. Results of the audit and system review should be documented and appropriate action implemented in relation to the findings.

Continuous Improvement: Establish a program of continuous improvement, which should include identification of potential weaknesses in the environmental management system, taking action to eliminate environmental burdens, identifying opportunities for improvement, and ensuring a continuous and dynamic orientation to a never-ending cycle of improvement.

SOUTH AFRICAN STANDARD: SABS-0251

In this section, we will limit our discussion to outlining only the salient features of the *South African Standard, SABS-0251: Environmental Management System*, and for that reason, the reader is advised to consult the original document for further details. However, it is quite interesting to note how different approaches to environmental management systems share a common set of system requirements and principles. This standard shares common system principles with ISO 9001.

SABS-0251: Environmental Management System—Salient Features

4.1 Management Responsibility

4.1.1 General

4.1.2 Environmental Policy

4.1.3 Organization

   4.1.3.1 Responsibility, Authority, and Resources

   4.1.3.2 Verification Resources, and Personnel

   4.1.3.3 Management Representative

   4.1.4 Environmental Management Review

4.2 Environmental Management System

4.3 Contract Review

4.4 Environmental Assessment and Planning

4.4.1 General

4.4.2 Statutory Requirements and Register

4.4.3 Environmental Assessment and Environmental Impact Register

4.4.4 Environmental Objectives and Targets

4.4.5 Environmental Management Program

4.5 Document and Data Control

4.5.1 General

4.5.2 Document Approval and Issue

4.5.3 Document Changes and Modifications

4.6 Purchasing

4.6.1 General

4.6.2 Suppliers and Subcontractors

4.6.3 Purchasing Data

4.6.4 Verification of Purchased Product

4.7 Control of Customer-Supplied Product

4.8 Identification and Traceability

4.9 Control of Processes and Activities

4.9.1 General

4.9.2 Control

4.10 Inspection and Testing

4.11 Control of Inspection, Measuring, and Test Equipment

4.12 Inspection and Test Status

4.13 Noncompliance

4.13.1 General

4.13.2 Control of Nonconforming Materials

4.13.2.1 Nonconforming Materials and Residual Materials

4.13.2.2 Disposal, with or without Concession, of Nonconforming Product or By-Product

4.14 Preventive and Corrective action

4.14.1 General

4.14.2 Preventive Action

4.14.3 Corrective Action

4.14.4 Emergencies

4.15 Handling, Storage, Packaging, and Delivery

4.16 Environmental Management Records

4.17 Environmental Audits

4.18 Training and Awareness

4.19 Servicing

4.20 Statistical Techniques

5

INTEGRATED QUALITY MANAGEMENT SYSTEM

In the previous chapters (and in Appendix B), we outline the subject of developing and implementing two systems aimed at improving quality—a Total Quality Management (TQM) system for products and services, and an Environmental Management System (EMS) for the environment. From this collective understanding of the two systems, we will now develop a set of guidelines for an integrated EMS/TQM system.

Organizations who have already implemented a suitable TQM system can use these guidelines to add the environmental management aspects required to establishing an effective EMS. On the other hand, for those organizations who are in their initial stages of quality system implementation, these guidelines provide substantial directions for implementing an integrated system that addresses both quality and environmental objectives.

QUALITY MANAGEMENT PRINCIPLES

We know there are a number of fundamental quality system management principles that are generically applicable to any business. At a minimum, any good quality management system will include: a dynamic customer orientation, participative management and effective leadership, integrated systems approach, total employee involvement, process management, system effectiveness, monitoring and evaluation, and continuous improvement focus.

Integrated EMS/TQM System: A Checklist

With our understanding of the necessary quality system imperatives, we can now develop the overall framework of the integrated EMS/TQM system. There are three broad components to this system (see Figure 5-1).

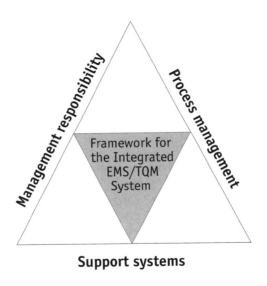

Figure 5-1. Three Components of an Integrated EMS/TQM System

Within each of the three components there is a checklist of system elements that are essential to developing and implementing an integrated EMS/TQM system.

Management Responsibility

- Management commitment to quality of products, services, and the environment

- Quality and environmental policy

- Quality and environmental objectives, goals, and targets

- Designation of appropriate quality and environmental responsibilities

- Identification of requisite resources for the implementation and maintenance of EMS/TQM system

- Defining appropriate leadership roles, responsibilities, and authority

- Establishment of process management teams for the EMS/TQM system

Process Management

- Establishment of training and development program

- Developing quality system and environmental system operating procedures

- Identifying customer quality needs

- Identifying stakeholder's environmental concerns

- Developing a profile of quality activities, and environmental aspects and impacts of all activities

- Implementing control systems for design, purchasing, process, production, calibration, handling, storage, delivery, and equipment maintenance

Support Systems

- Developing process monitoring and measurement system

- Establishing procedure for the identification of nonconformances and taking of corrective and preventive action

- Establishing performance measurement and evaluation system

- Establishing a continuous improvement program

INTEGRATED EMS/TQM SYSTEM: A ROAD MAP

Based on the checklists above and in line with the TQM and EMS implementation road maps found in Chapter 3 and Appendix B, we will now outline a ten phase road map for developing and implementing an integrated EMS/TQM system (see Figure 5-2).

Phase 1: Management Readiness

- Develop quality policy and environmental policy.

- Establish organization's vision, mission, core values, beliefs, and guiding principles for the maintenance and improvement of quality of products, services, and the environmental aspects of activities.

- Identify tangible management commitment to quality and environmental policy and communicate the same to employees at all levels of the organization.

- Arrange for a management orientation and awareness session on all aspects of quality.

- Ensure that all key managers and supervisors are committed to the quality and environmental policy and are accountable for the same.

- Appoint an EMS/TQM coordinator responsible for the implementation and maintenance of the EMS/TQM system.

- Establish a steering committee to oversee the effective functioning of the EMS/TQM system.

Phase 2: Customer-Supplier Partnering

- Highlight activities to identify dynamic customer focus.

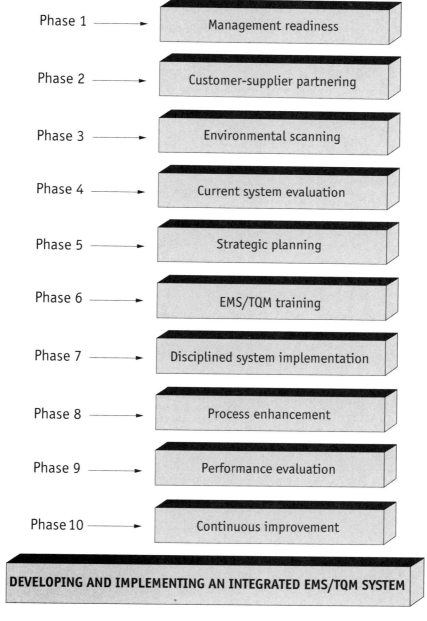

Figure 5-2. Integrated EMS/TQM System: A Road Map

- Ensure that adequate customer and stakeholder feedback is obtained regarding the quality aspects and concerns of products, services, and the environment.

- Ensure that the quality and environmental policy and objectives are commensurate with the needs and expectations of the customers.

- Ensure that quality and environmental objectives, goals, and targets are continuously reviewed and revised vis-à-vis input from the shareholders.

- Identify and ensure compliance to the legal and regulatory requirements concerning the impact of activities on the environment.

- Establish a close working relationship with suppliers and subcontractors. Make sure that the suppliers clearly understand the company's requirements for quality and environmental aspects of products and services acquired. Assist the suppliers in utilizing practices and raw materials that minimize environmental impacts.

- Establish practices to highlight constancy and trust with both the suppliers and the customers.

Phase 3: Environmental Scanning

- Develop a profile of marketplace needs and expectations in relation to the quality of your product and service offerings.

- Develop a profile of environmental concerns of the customers, regulators, and of society in general in relation to your activities, products, and services.

- Identify what your clients expect in terms of quality, technological innovativeness, price, and environmental safety.

- Identify what design aspects can be introduced to produce products and services that would be competitively superior in quality and efficient in environmental aspects.

- Assess your own market share. Compare it with that of your competitors and identify opportunities for growth and improvement.

- Implement a continuous improvement strategy for developing new and innovative product lines or designs for your current and potential customers.

- Develop a profile of the organization's activities throughout the various life-cycle stages and identify their environmental aspects.

- Evaluate the environmental effects of these activities.

- Identify what changes in processes, practices, methodologies, or technologies can be instituted to minimize or eliminate the environmental loadings on the environment.

- Identify the environmental burdens associated with the materials and services acquired from your suppliers and subcontractors and establish procedures to work with your suppliers to minimize the environmental impacts of those materials.

- Identify the needs, requirements, and expectations of your employees to enhance the quality of work life, and to improve the quality of products, services, and the environment.

Phase 4: Current System Evaluation

- Develop a master list of activities of the organization relating to quality improvement and environmental safety.

- Develop a profile of the environmental aspects of activities, e.g., emissions to air, releases of water, contamination, waste management.

- Develop a profile of all current initiatives and programs of the organization for the improvement of quality and environmental aspects of products and services.

- Conduct a gap-analysis to evaluate strengths and weaknesses of the current systems in relation to the needs and expectations identified through environmental scanning.

- Evaluate the suitability of current practices in terms of their:

  – infrastructural suitability for quality improvement and environmental safety

  – adequacy of requisite resources: physical, financial, human, technical

  – adequacy of proper procedures, processes, and methodologies

  – focus for customer and stakeholder satisfaction

  – compliance with the environmental regulations and codes of conduct

- Identify the status of system documentation for quality and environmental aspects, and evaluate its operational suitability and adequacy.

- Identify the suitability of training and development activities relating to general as well as specific aspects of product/service quality and environmental burdens.

Phase 5: Strategic Planning

- Establish and document quality and environmental objectives, goals, and targets commensurate with quality and environmental policy.

- Communicate the objectives to all levels of the organization.

- The goals and targets should be measurable.

- The objectives must be commensurate with customer, legislative, and regulatory requirements.

- Develop a suitable EMS/TQM model that is:

  - commensurate with the current systems, procedures, and practices

  - user-friendly and sustainable

  - people-developed and people-empowered

- Develop a master EMS/TQM implementation plan with time schedules, responsibilities, check-points, review procedures, performance evaluation framework, and continuous improvement program.

- Assign appropriate responsibilities and authority for the implementation, operation, and maintenance of the system. Establish accountability framework. Document and communicate responsibilities and authorities to everyone in the organization.

- Identify and allocate appropriate resources: human, physical, technological, and financial.

- Establish process improvement and process management teams.

- Involve and empower people at all levels of the organization to implement, maintain, and improve the quality of products/services and the environment.

- Develop and maintain EMS/TQM system documentation hierarchy: quality manual, quality system procedures, standard operating procedures, work instructions, etc.

Phase 6: EMS/TQM Training

- Identify training needs relating to quality and environment aspects.

- Establish an EMS/TQM training schedule for all employees.

- Provide the requisite training appropriate to all levels of the organization.

- Conduct continuous awareness sessions throughout the organization on the control, maintenance, and improvement of EMS/TQM system.

- Personnel whose activities have been designated as having a significant impact on quality and environmental aspects should be given adequate specialized training. The selection of these personnel should be on the basis of their knowledge, training, and experience.

- Adequate training should be provided on all aspects of quality and environment including, as a minimum, the following:

  – process management

  – process improvement

  – use of procedures

– control of procedures

– auditing

– process/equipment capability

– performance evaluation

Phase 7: Disciplined System Implementation

- Establish, maintain, and utilize documented procedures (Standard Operating Procedures) in all areas of activity. These procedures must:

  – adequately address all aspects of product/service and environmental quality

  – be available at all locations, as required

  – be accessible to everyone who needs them

  – be reviewed and revised regularly

  – be properly controlled

  – be managed by designated responsibility centers and contact points

- Establish critical control points at all areas of activity, such as: material acquisition, process, design, production, customer service, and waste management.

- Establish a master control system for all documents and data pertaining to the quality system and environmental aspects.

- Establish adequate environmental acceptance criteria for operational control of procedures.

- Establish environmental emergency response procedures.

- Establish and maintain training and development programs.

- Establish EMS/TQM system audit program.

- Develop a suitable EMS/TQM record control system.

- Establish mechanisms for identifying new initiatives, development projects, or changes in technology to improve quality and design of products, services, and the environment.

- Establish procedures for process and equipment capability studies.

- Establish employee motivational programs.

Phase 8: Process Enhancement

- Establish procedures to measure, monitor, and analyze key product/environmental quality characteristics at all process control points.

- Establish procedures for the review, revision, and control of standard operating procedures and other EMS/TQM related procedures.

- Establish process improvement teams to evaluate process outputs and to recommend solutions and initiatives for process enhancement.

- Develop suitable work instruction manuals for operational processes and utilize these for continuous training and retraining.

- Make use of suitable process improvement tools and statistical process control methods.

Phase 9: Performance Evaluation

- Establish procedures for conducting performance reviews of the EMS/TQM system at regular intervals.

- Quantify and document performance evaluation findings.

- Conduct regular system audits.

- Document improvements made with respect to quality and environmental issues.

- Establish procedures for the identification of nonconformances and taking of corrective and preventive action at each operational unit.

- Revise operating procedures in relation to the corrective and preventive actions.

- Conduct management reviews of the EMS/TQM system.

- Identify activities and initiatives for the design and acquisition of new technologies for upgrading the system.

Phase 10: Continuous Improvement

- Make continuous improvement the focal point of all activity centers.

- Establish measurable continuous improvement programs.

- Set up a framework for identifying improvement opportunities.

- Establish strategic initiatives and allocate requisite resources for the improvement programs.

- Develop a continuous training and development program.

- Focus on never-ending cycle of continuous improvement.

PART 2

ENVIRONMENTAL QUALITY CERTIFICATION

6

EMS CERTIFICATION—
ISO 14001

The fundamental mission of most organizations is to achieve success—produce more with less, expand market share, and realize growth and profit. But this is the old paradigm of success. Today the new vision of *competitive success* in the year 2000 and beyond requires equal attention to the quality management of products and services, and the quality management of environmental impacts. Today it is very important that organizations look beyond the old limited model of success and acknowledge a responsibility towards the customers and society to provide high quality goods and services that, along with their associated activities, do not impose any environmental burdens on society.

A well-proven way to fulfill this responsibility is to establish and maintain effective quality management systems and to commit to continuous improvement. Part I was devoted to providing good manufacturing practices, principles, and guidelines for establishing such systems. The next question that faces the organization relates to providing tangible evidence of successfully implemented and operating quality management systems. This function is adequately performed by the quality system certification program. To achieve the certification accreditation, companies have to establish a suitable quality system, and demonstrate to a third-party registrar that the system has been effectively implemented and consistently maintained, as per the requirements of the stipulated standard.

The most viable option for quality system certification pertaining to products and services is certification to ISO 9000 quality system standards. There are currently three options an organization can exercise for environmental management system certification. The organization can self-regulate their functions in relation to the existing environmental management protocols or regulatory requirements; seek certification to the environmental management system specification of the British Standard: BS 7750; or prepare for EMS certification to the international standard, ISO 14001.

For quality system certification, most organizations have either achieved or are in the process of achieving the ISO 9000 certification status. This chapter provides guidelines for certification to the ISO 14001 standards for environmental quality management systems. Organizations that need either the environmental management system certification or a combined certification for the environmental quality as well as product/service quality management systems, can either utilize the guidelines appended in this chapter or the detailed integrated set of guidelines presented in the next chapter. Before proceeding to describe the process of EMS certification, we shall present the requisite specifications outlined in the draft international standard, *ISO 14001: Environmental Management Systems—Specifications with Guidance for Use.*

EMS SPECIFICATIONS: ISO 14001

There is a clear distinction between a guidance standard and a compliance standard. The focus of ISO 14004, as outlined in the previous sections, is on providing guidance for implementing and maintaining an environmental management system. ISO 14001, on the other hand, provides a prescriptive set of specifications that may be objectively audited for certification or self-declaration purposes. As such, ISO 14001 will serve as the compliance standard for EMS certification. Indeed, the system elements appended in ISO 14001 can also be utilized to develop an environmental management system.

Table 6-1 presents the overall format of ISO 14001 specifications and a detailed description of each element.

Table 6-1. ISO 14001—Environmental Management System

EMS Specifications	Details
4.0 General	• Establish and maintain an effective environment management system to improve the quality of the environment.
4.1 Environmental Policy	• Define and document environmental policy. The policy should: – Be relevant to the nature, scale, and environmental impacts of activities, products, and services – Include a commitment to continual improvement – Be in compliance with environmental legislative and regulatory requirements – Provide framework for establishing environmental objectives and targets – Be documented, implemented, maintained, and communicated to all employees – Be made available to the public
4.2 Planning	
4.2.1 Environmental Aspects	• Establish and maintain procedures to identify and control the environmental aspects of activities, products, and services which can have significant impacts on the environment. This process should consider: – emission to air – releases of water – releases to or contamination of land – use of raw materials and natural resources – other local environmental issues • The process requires the identification of only the most significant environmental aspects of products and services, and does not require a detailed life cycle assessment. The nature and type of control

Table 6-1. ISO 14001—Environmental Management System *(cont'd)*

EMS Specifications	Details
	may vary significantly, e.g., altering just a single input material or altering the product design.
4.2.2 Legal and Other Requirements	• Establish and maintain procedures to ensure compliance with legislative and regulatory requirements, such as: – industry code of practices – agreement with public authorities – nonregulatory guidelines
4.2.3 Objectives and Targets	• Establish and maintain documented environmental objectives and targets, at all levels within the organization, that are: – commensurate with the legislative and regulatory requirements; environmental aspects; and the technological, financial, operational, and business requirements of the organization and interested parties – consistent with the environmental policy • Objectives should be specific and targets should be measurable wherever possible, and where appropriate preventive measures should be taken into account.
4.2.4 Environmental Management Program(s)	• Establish and maintain a program for achieving the objectives and targets. It should include: – means and time frame for achieving objectives and targets – designation of appropriate responsibilities – environmental review process encompassing the planning, design, production, marketing, and disposal of both current and new products, services, and activities
4.3 Implementation and Operation	
4.3.1 Structure and Responsibility	• Define, document, and communicate roles and responsibilities and accountability for the management of environmental issues.

EMS Specifications	Details
	• Provide requisite personnel, technical, and financial resources for the implementation and verification of environmental policies and objectives.
	• Appoint a management system representative to ensure: – effective system implementation – reporting on the performance of the system to management for review and improvement
4.3.2 Training, Awareness, and Competence	• Identify training needs and provide appropriate training to personnel whose work may create a significant impact upon the environment.
	• Employees, at all levels, should be trained to be aware of: – importance of conformance to the policy and procedures of the EMS – significant environmental aspects of their work activities and the environmental benefits of improved personal performance – their roles and responsibilities in achieving conformance with the environmental policy and procedures, and with the requirements of the system – potential consequences of departure from procedures
	• Personnel for specific assigned tasks should be selected on the basis of their training, knowledge, and experience.
4.3.3 Communications	• Establish and maintain suitable means of communication for receiving and relaying relevant information and data internally and externally.
4.3.4 Environmental Management System Documentation	• Establish and maintain documented information system for the EMS.
4.3.5 Document Control	• Establish and maintain a document

Table 6-1. ISO 14001—Environmental Management System *(cont'd)*

EMS Specifications	Details
	control system to ensure that all requisite documents are: – properly identified – periodically reviewed, revised, and approved by authorized personnel prior to use – available at all appropriate locations – promptly removed when they become obsolete – legible, readily identifiable, and effectively maintained
4.3.6 Operational Control	• Ensure that the EMS activities are carried out under controlled conditions by preparing appropriate documented procedures and establishing capability and acceptance criteria of operating controls. • Establish procedures for controlling significant environmental aspects of products and services and also communicating these procedures and requirements to the suppliers and contractors.
4.3.7 Emergency Preparedness and Response	• Establish and maintain procedures for responding to emergency situations. Revise and update these procedures as appropriate. • Emergency response procedures should also be periodically tested whenever applicable.
4.4 Checking and Corrective Action	
4.4.1 Monitoring and Measurement	• Establish procedures to monitor key process characteristics. • Establish procedures for calibration of monitoring equipment and maintenance of calibration records. • Establish procedures for evaluating compliance with relevant environmental legislation and regulations.

EMS Specifications	Details
4.4.2 Nonconformance and Corrective and Preventive Action	• Define and document responsibilities for the identification of nonconformances and taking of corrective and preventive action. • Establish procedures for the identification of nonconformance and initiation of corrective and preventive action. These actions should be commensurate with the magnitude of risk encountered. • Implement and record any changes in the documented procedures resulting from corrective and preventive action.
4.4.3 Records	• Establish procedures for the identification, maintenance, and disposition of requisite EMS records. • Records should be: legible, identifiable, traceable, properly handled and stored, and readily retrievable.
4.4.4 Environmental Management System Audit	• Establish and maintain a program and procedures for audit to ensure that: – the EMS conforms to the requirements of this standard – the system has been effectively implemented and maintained • Results of the audit should be presented to management for review and action.
4.5 Management Review	• Management should review the EMS at regular intervals to ensure its continuing suitability and effectiveness. Records of reviews shall be maintained.

EMS CERTIFICATION PROCESS: ISO 14001

The basic premise of ISO 14001 can be stated as follows:

- Organizations should develop an environmental policy with objectives and targets commensurate with the environmental aspects of their activities and their impact on the environment.

- An environmental management system should be established to ensure conformance with the stated policies and objectives.

- The organization should be able to demonstrate conformance to stated environmental policies and principles either through certification to ISO 14001 or through self-declaration.

- The environmental quality system should be effectively maintained through a program of continual improvement.

The overall process of environmental quality management system implementation and certification to ISO 14001 can be compartmentalized into six distinct phases (see Figure 6-1).

Phase 1: Awareness

No system can be effectively implemented unless the people responsible for the system implementation are totally aware of the requirements of the system. How much and in what detail the awareness should be provided at various levels in the organization is a matter of judgment for the management. A typical action plan should involve the following:

- The quality manager/ISO coordinator and a few key personnel should undertake the initial training in EMS implementation from some reputable external source.

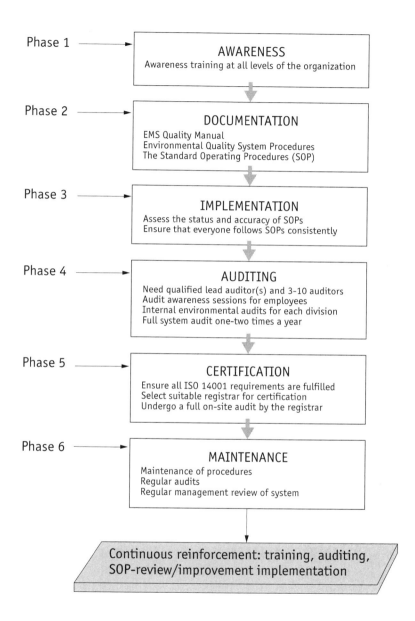

Figure 6-1. Implementation and Certification to ISO 14001

- A good consultant can be hired to provide the following in-house training:

  – a short awareness session on EMS principles to the executive management

  – a one or two-day session on the detailed content of ISO 14001 and EMS implementation to various groups from the ranks of managers and supervisors

- The managers and supervisors who have been trained should now bear the responsibility of providing continuous short training sessions to all employees at the working levels.

- It is imperative to ensure that everyone in the organization receives some training on EMS principles commensurate with their needs and requirements.

Phase 2: Documentation

For the effective functioning of the system, the system must be properly documented, maintained, and consistently followed by everyone in the organization. Generally an organization will need to establish the following three documents:

- EMS Quality Manual

- Environmental Quality System Procedures

- The Standard Operating Procedures (SOPs)

EMS Quality Manual: The EMS Quality Manual is generic in nature. It addresses the system requirements of ISO 14001 while outlining the environmental quality system and procedures of the company. The manual also outlines the environmental policy, objectives, and responsibility framework of the organization.

Environmental Quality System Procedures: These procedures are really an expanded version of the quality manual. While the manual confirms the existence of procedures, the system procedures detail how these procedures are implemented and carried out.

Standard Operating Procedures (SOPs): SOPs are, typically, the operating procedures associated with all the functional and manufacturing activities of the company. They outline environmental quality management aspects of these operational activities. They should be updated to accommodate the environmental concerns inherent at the working level. These procedures have to have a linkage to the system level procedures for all the requisite quality system elements.

Further guidance on how to develop these three documents is provided in Chapter 7.

Phase 3: Implementation

Implementation is the most crucial phase in the certification process. Although system implementation takes time, it is by no means intractable. The following action plan is recommended:

- Assess the status of Standard Operating Procedures (SOPs) to ensure that each system and division has well-documented and established procedures and everyone follows them consistently.

- Each of the operating procedures should include the various requisite aspects of environmental quality management.

- These SOPs should address the requisite system aspects, such as:

  – procedures for making changes and for approval of documents

  – system for controlling all documentation

  – system for identification of nonconformances and taking of corrective and preventive action

– suitable means of keeping records

– system for identification and traceability

– audit and review procedures for system effectiveness

– emergency response procedures

Phase 4: Auditing

The success of a system is totally dependent on the effectiveness of the audit function. A detailed discussion on the subject of environmental auditing is covered in Chapter 8; however, the following requirements should be noted:

- Depending on the size of the organization, there is a need to have at least one or two qualified Lead Auditors and between three to ten auditors. However, more trained auditors means better system maintenance.

- Several audit awareness sessions should be conducted for the employees by qualified auditors to explain the process of internal and external auditing.

- Numerous internal environmental audits need to be conducted for each division prior to the first full audit by the external certification body, and this exercise is continued throughout the certification period.

- Full system audit should be conducted at least once or twice each year.

- Continuous environmental audit training and retraining should be an integral part of the system.

Phase 5: Certification

Once the environmental system has been adequately documented, implemented, and internally audited for conformance to ISO 14001 requirements, it is time for a certification audit by the registrar. The following items should be considered:

- Ensure that all operating procedures are effectively implemented and functionally maintained.

- Ensure that all the ISO 14001 system requirements are fulfilled.

- Provide adequate awareness and training to employees on the overall external audit process.

- Select a suitable registrar for certification to ISO 14001.

- Undergo a full on-site audit by the registrar.

- Remove any deficiencies identified by the registrar.

- Achieve EMS certification.

- Maintain the EMS.

Phase 6: Maintenance

After a certificate has been awarded, the next most important task is the maintenance of the environmental quality system. A poorly maintained system will have a high chance of failing a follow-up audit, and a company can easily lose its certificate, and hence its credibility, at anytime during or after the certification period. The basic activities associated with system maintenance involve:

- Maintenance of procedures

- Regular audits

- Continuous training

- Effective control of nonconformances

- Implementation of preventative measures

- Regular management review of the system

The system effectiveness can be measured by continuously analyzing the following information:

- Rate of nonconformances

- Nature and type of defects arising in the system

- Audit reports

- System nonconformances

- Customer complaints

- Rate of rejects and returns

- Rate of environmental emergencies

ISO 14001 IMPLEMENTATION ROAD MAP

In line with the activities associated with the six implementation phases, we can now outline an overall sequential plan for the certification process. The following order of events is recommended for the implementation of an ISO 14001 environmental quality management system:

1. Top management undergoes ISO 14001 orientation.

2. The quality manager and the steering group develop a strategic implementation plan.

3. A decision is made whether to hire an outside consultant to facilitate the implementation process.

4. ISO 14001 training seminars are provided to the managers and supervisors.

5. The quality manager starts the development of the EMS quality manual.

6. The quality manager and key technical managers formulate process management teams to start the development of

system level procedures and the standard operating procedures. The quality manager provides the leadership role in this exercise.

7. Regular ISO 14001 awareness sessions are given to employees at all levels in the organization.

8. The steering group undertakes the responsibility of developing environmental policy, objectives, targets, and mission statement.

9. Key personnel are identified to take responsibility for various important aspects of the environmental system, e.g., training, document control, audit, emergency response.

10. All environmental quality system documentation is reviewed by key personnel.

11. The quality manager and another key person undergo lead auditor training.

12. Audit training is provided to a select group of people who will manage the environmental audit program.

13. All operational managers are to identify environmental aspects and impacts of their activities.

14. EMS implementation begins.

15. Procedures and processes are put into place.

16. Critical control points are established to monitor the system.

17. Routine internal audits are performed.

18. System deficiencies are identified and corrected.

19. Contact is established with the chosen registrar for the conduct of precertification audit, if desired.

20. A final evaluation is made to ensure that the system is functioning effectively.

21. The registrar performs the complete on-site certification audit.

22. Deficiencies identified through external audit, if any, are corrected.

23. Company is awarded the certificate.

24. System is maintained and assessed continuously for compliance with surveillance audits conducted by the Registrar during the span of the certification period.

ISO 14001: IMPLEMENTATION CHECKLIST

In this section there is a user-friendly checklist of ISO 14001 requirements for a hands-on implementation of the environmental quality management system. This checklist can also be beneficial in developing the EMS quality manual that would accurately accommodate the ISO 14001 requirements. The user may modify this checklist and develop their own commensurate with their needs.

Table 6-2. ISO 14001 Implementation User-Friendly Checklist

Legend

🖙🖙🖙 Need Action

🖙🖙 Completed Action

🖙 N/A

SYSTEM ELEMENTS	🖙🖙🖙	🖙🖙	🖙
4. Environmental Management System			
4.0 General			
• Establish and maintain an environmental management system.			
4.1 Environmental Policy			
• Define and document the environmental policy. Ensure that the policy:			
– is understood, implemented, and maintained at all levels of the organization			
– is commensurate with the nature and magnitude of environmental impacts of activities			
– identifies commitment to continuous improvement and elimination of environmental burdens			
– is consistent with the relevant environmental legislation and regulations			
– generates a framework for continuous review of environmental objectives and targets			
– is made publicly available			
4.2 Planning			
4.2.1 Environmental Aspects			
• Establish and maintain procedures to identify the environmental aspects of activities throughout their life cycle stages.			
• Evaluate and document the environmental impact of these activities.			
• Ensure that the environmental objectives adequately address and accommodate the environmental impacts of activities, products, and services.			
• Ensure that information regarding the environmental aspects of activities and their impact is			

Table 6-2. ISO 14001 Implementation User-Friendly Checklist *(cont'd)*

SYSTEM ELEMENTS	☞ ☞ ☞ ☞	☞ ☞	☞
continuously reviewed, updated, and communicated to everyone concerned.			
4.2.2 Legal and Other Requirements			
• Establish and maintain procedures for identifying legal and other regulatory requirements concerning the environmental aspects of activities, products and services.			
4.2.3 Objectives and Targets			
• Establish environmental objectives and targets at each relevant level and function within the organization.			
• Ensure that the objectives and targets are consistent with the environmental policy and commitment.			
• Continuously review the objectives vis-à-vis: the legal and regulatory requirements; nature and impact of environmental aspects of activities; technological, financial and business requirements; and needs and concerns of the stakeholders.			
4.2.4 Environmental Management Program(s)			
• Establish and maintain environmental management program, including:			
– designated responsibilities at all levels of the organization for achieving objectives and targets			
– goals, schedules, and timeframes for the achievement of objectives			
• The program should be extended to include new developments and new or modified activities, products, and services.			
4.3 Implementation and Operation			
4.3.1 Structure and Responsibility			
• Define, document, and assign appropriate roles, responsibilities, and authorities for all aspects of the EMS.			
• Provide adequate human, technological, and financial resources for the effective implementation, control, and functioning of the EMS.			

SYSTEM ELEMENTS	☞☞☞	☞☞	☞

• Appoint a management representative, with defined responsibility and authority, to:

   – ensure that the EMS is established, implemented, and maintained in accordance with the requirements of the ISO 14001

   – report to the management on the performance of the system for review and improvement

4.3.2 Training, Awareness, and Competence

• Identify training needs and provide appropriate training to all personnel, commensurate with the nature of their activity and the significance of its impact on the environment.

• Establish procedures to ensure that employees at all levels of the organization are aware of:

   – the importance of conformance to environmental procedures, policy, and EMS requirements

   – the environmental impacts of their activities

   – the environmental benefits that would accrue from improved personal performance

   – their roles and responsibilities in achieving conformance to the environmental policy, procedures, EMS requirements, and the emergency response requirements

   – the potential consequences of departure from the specified operating procedures

• Personnel whose work involves activities which may have a significant impact on the environment should be selected on the basis of appropriate education, training, and experience.

4.3.3 Communication

• Establish and maintain procedures for:

   – organizational and technical interface and communication between the various levels and functions within the organization

   – receiving, documenting, and responding to relevant communication from external stakeholders on the

Table 6-2. ISO 14001 Implementation User-Friendly Checklist *(cont'd)*

SYSTEM ELEMENTS	☞ ☞ ☞	☞ ☞	☞
environmental aspects and the environmental management system			
• Maintain records of decisions on the organization's significant environmental aspects and their communication to the external stakeholders.			
4.3.4 Environmental Management System Documentation			
• Establish and maintain information, in paper or electronic form, to:			
– outline the core EMS elements, their requirements and interaction			
– provide direction to appropriate relevant documentation			
4.3.5 Document Control			
• Establish and maintain an effective document control system. Ensure that the documents are:			
– available at all requisite locations			
– periodically reviewed, revised as necessary, and approved for adequacy by authorized personnel			
– promptly removed from all points of issue or use when they become obsolete—obsolete documents can be retained for legal and/or knowledge preservation purposes as long as they are suitably identified.			
– legible, readily identifiable, maintained in an orderly manner, and retained for a specified period of time			
– properly stored and easily retrievable			
• Establish and maintain procedures identifying appropriate responsibilities and authorities for the creation, review and revision of the various types of environmental documents.			
4.3.6 Operational Control			
• Identify the operations and activities that exert significant environmental impact, and establish control procedures to ensure that these activities are carried out under specified conditions. Control procedures could include:			
– establishing and maintaining documented procedures			

SYSTEM ELEMENTS	☞☞☞☞	☞☞	☞

    – stipulating operating criteria in the procedures

    – partnering with suppliers and subcontractors to ensure that they clearly understand the organization's environmental concerns and requirements.

4.3.7 Emergency Preparedness and Response

- Establish and maintain procedures to respond to environmental emergencies.

- Continually review and revise the emergency response procedures.

- Periodically test these procedures, where practicable.

4.4 Checking and Corrective Action

4.4.1 Monitoring and Measurement

- Establish and maintain procedures to continuously monitor and measure the key environmental characteristics. Record the information to trace the performance and conformance of operational controls with the environmental objectives and targets.

- Establish procedures for calibrating the monitoring equipment. Maintain records of calibration.

- Establish procedures for evaluating compliance with the relevant environmental legislation and regulations.

4.4.2 Nonconformance and Corrective and Preventive Action

- Establish and maintain procedures for defining responsibility and authority for handling nonconformances, and for taking appropriate corrective and preventive action.

- The corrective or preventive action should match the magnitude of risk encountered with the environmental impact.

- Implement and record any changes in the documented procedures resulting from the corrective and preventive actions.

4.4.3 Records

- Establish and maintain procedures for the identification, maintenance, and disposition of environmental records.

Table 6-2. ISO 14001 Implementation User-Friendly Checklist *(cont'd)*

SYSTEM ELEMENTS	☞ ☞ ☞	☞ ☞	☞
• Maintain training records as well as records of the results of audits and reviews.			
• Records should be legible, identifiable, and traceable to the activity.			
• Records should be stored and maintained in such a way that they are easily retrievable and protected against damage, deterioration, or loss.			
• Retention times of records should be established and recorded.			
• Records should be maintained to demonstrate conformance to the requirements of ISO 14001.			
4.4.4 Environmental Management System Audit			
• Establish and maintain a program and procedures for carrying out regular audits of the EMS.			
• The audits should determine whether the EMS has been properly implemented and maintained and conforms to the planned arrangements as per the requirements of ISO 14001.			
• Results of the audit should be reported to the management for review and improvement of the system.			
• The frequency and intensity of audits should be commensurate with the environmental importance of the activity concerned.			
4.5 Management Review			
• Management should review the EMS at regular intervals to ensure its continuing suitability, adequacy, and effectiveness.			
• Records of the review should be maintained.			
• The management review should consider the need for changes to policy, objectives and other aspects of the EMS in relation to the audit results, changing and evolving technologies, or any other aspect relating to continual improvement.			

7

INTEGRATED QUALITY SYSTEM CERTIFICATION

Throughout this book we have been emphasizing that if an organization has a need to implement both the EMS and TQM systems and achieve certification to both the ISO 9001 (or ISO 9002) and ISO 14001 standards, they should try to prepare their documentation and implementation for the combined systems rather than preparing two sets of quality system documentation. By doing so, they can achieve both certifications at the same time, and realize tremendous savings in effort, time, and money.

This chapter collates the information presented previously on quality system management and certification into hands-on checklists and road maps to prepare the organization for certification to the integrated set of requirements for ISO 9001 and ISO 14001 (as and when the facilities and options for such an integrated certification are available). The guidelines can be easily used for certification to ISO 9001, ISO 14001, or the integrated ISO 9001 and ISO 14001 system certification.

Before proceeding, we will reiterate some of the logistics of the matter. So far we have elucidated guidelines for developing and implementing a Total Quality Management (TQM) system; an Environmental Management System (EMS); an integrated EMS/TQM system; and the process of EMS certification to ISO 14001.

We have also identified the available system certifications for those organizations that wish to achieve certification. They are: the ISO 9000 quality system standards, and for the environmental management system, the British Standard, BS 7750 and the international standard, ISO 14001.

THE ISO 9000 SERIES

The quality system standards in the ISO 9000 series have been developed by the technical committee, ISO/TC 176, of the International Organization for Standardization (ISO). The two basic objectives of the committee are to provide organizations a set of *guidance standards* for quality management systems and a set of *conformance standards* for accreditation and certification of quality systems.

> ***Guidance standards:*** Organizations can utilize these standards to establish, implement and maintain quality management systems. In this category, the committee has also developed several other helpful standards on supporting technologies to assist organizations in the improvement and sustainability of quality systems. The committee continues to develop other relevant standards as per the marketplace needs. The standards developed to date under the guidance category are listed in Table 7-1.

Table 7-1. ISO/TC 176—Quality System Standards

ISO 9000 Quality Management and Quality Assurance Standards

ISO 9000-1	Part 1	Guidelines for Selection and Use
ISO 9000-2	Part 2	Generic Guidelines for Application of ISO 9001, ISO 9002, and ISO 9003
ISO 9000-3	Part 3	Guidelines for the Application of ISO 9001 to the Development, Supply, and Maintenance of Software
ISO 9000-4	Part 4	Application for Dependability Management

ISO 9004 Quality Management and Quality System Elements

ISO 9004-1	Part 1	Guidelines
ISO 9004-2	Part 2	Guidelines for Services
ISO 9004-3	Part 3	Guidelines for Processed Materials
ISO 9004-4	Part 4	Guidelines for Quality Improvement
ISO 9004-6	Part 6	Guide to Quality Assurance for Project Management
ISO 9004-7	Part 7	Guidelines for Configuration Management
ISO 9004-8	Part 8	Guidelines on Quality Principles and Their Application to Management Practices
ISO 10005		Quality Management: Guidelines for Quality Plans

ISO 10011 Guidelines for Auditing Quality Systems

ISO-10011-1	Part 1	Auditing
ISO-10011-2	Part 2	Qualification Criteria for Quality System Auditors
ISO-10011-3	Part 3	Management of Audit Programs

ISO 10012 Quality Assurance Requirements for Measuring Equipment

ISO-10012-1	Part 1	Metrological Qualification System for Measuring Equipment
ISO-10012-2	Part 2	Measurement Assurance

ISO 10013 Preparation of Quality Manual

ISO 10014 Economics of Quality

ISO 10015 Continuing Education and Training Guidelines

ISO 10016 Records of Product Inspection and Test Guidelines for Preparation of the Results and their Conformance

Conformance standards: Organizations can use these standards to seek accreditation and certification of their quality systems. This category includes the three-tier conformance model: ISO 9001, ISO 9002, and ISO 9003 as presented in Table 7-2.

Table 7-2. ISO 9000 Series (1994)—Quality System Standards

	Conformance Standards
ISO 9001	Quality Systems–Model for Quality Assurance in Design/Development, Production, Installation, and Servicing
ISO 9002	Quality Systems–Model for Quality Assurance in Production, Installation, and Servicing
ISO 9003	Quality Systems–Model for Quality Assurance in Final Inspection and Test

The guidance standards are descriptive documents that are only advisory in nature, providing guidelines for establishing, monitoring, and improving quality systems. The conformance standards, on the other hand, specify prescriptive requirements which can be used for auditing the quality system for certification.

The first set of standards in the ISO 9000 series was introduced in 1987. As per ISO directives, all ISO standards are reviewed and revised every five years. Tables 7-1 and 7-2 contain the revised, 1994, versions of most of the standards. The next revision of ISO 9001, 9002, and 9003, known in the committee circles as revision #2, is expected to be available by 1997. It is anticipated that the second revision of these conformance standards will exhibit some major structural and format changes. The intent of these changes is to make the standards more user-friendly and accentuate the importance of quality management.

THREE-TIER ISO 9000 CONFORMANCE MODEL

The three-tier model, ISO 9001, 9002, and 9003, constitutes the actual series of conformance standards for a third-party certification. Each model is complete, independent, and contains a set of quality system requirements. ISO 9001 is the most comprehensive while ISO 9003 is the least comprehensive. It is for each company to select the appropriate model according to their quality system needs. A comparative schematic of the quality system elements of the three levels is shown in Table 7-3.

The type of information contained in these three models is as follows:

Model 1: ISO 9001

This model is for use when conformance to specified requirements is to be assured by the supplier throughout the whole cycle from design, production, and installation to servicing. ISO 9001 consists of twenty required quality system elements and represents the fullest and most stringent requirements for a quality system as outlined in the guidance standard ISO 9004-1. ISO 9001 is applicable to organizations such as engineering and construction firms and manufacturers or even service companies, who design, develop, produce, install and service products.

Model 2: ISO 9002

The ISO 9002 model is identical to and equally as stringent as ISO 9001 except that of the twenty required quality system elements it does not include Design Control. This model is applicable to organizations that do not have any design requirements and is for use when conformance to specified requirements are to be assured during production, installation and servicing. It is particularly suited to process industries (food, chemical, pharmaceutical, etc.) where the specific requirements for the product are stated in terms of an already established design or specification.

Table 7-3. ISO 9000 Series (1994)—International Quality System Standards

Clause #	ISO 9001	ISO 9002	ISO 9003
4.1	Management Responsibility	Management Responsibility	Management Responsibility
4.2	Quality System	Quality System	Quality System
4.3	Contract Review	Contract Review	Contract Review
4.4	Design Control		
4.5	Document and Data control	Document and Data Control	Document and Data Control
4.6	Purchasing	Purchasing	
4.7	Control of Customer Supplied Product	Control of Customer Supplied Product	Control of Customer Supplied Product
4.8	Product Identification and Traceability	Product Identification and Traceability	Product Identification and Traceability
4.9	Process Control	Process Control	
4.10	Inspection and Testing	Inspection and Testing	Inspection and Testing
4.11	Control of Inspection, Measuring and Test Equipment	Control of Inspection, Measuring and Test Equipment	Control of Inspection, Measuring and Test Equipment
4.12	Inspection and Test Status	Inspection and Test Status	Inspection and Test Status
4.13	Control of Nonconforming Product	Control of Nonconforming Product	Control of Nonconforming Product
4.14	Corrective and Preventive Action	Corrective and Preventive Action	Corrective and Preventive Action
4.15	Handling, Storage, Packaging, Preservation and Delivery	Handling, Storage, Packaging Preservation and Delivery	Handling, Storage, Packaging Preservation and Delivery
4.16	Control of Quality Records	Control of Quality Records	Control of Quality Records
4.17	Internal Quality Audits	Internal Quality Audits	Internal Quality Audits
4.18	Training	Training	Training
4.19	Servicing	Servicing	
4.20	Statistical Techniques	Statistical Techniques	Statistical Techniques

Model 3: ISO 9003

The ISO 9003 model relates only to those elements concerning final inspection and test. It has the least number of requirements of the three models. It is suitable for small shops, divisions within an organization, laboratories, or equipment distributors that inspect and test supplied products.

CERTIFICATION PROCESS: ISO 9001 AND ISO 14001

By following the format and framework of ISO 9001 as the basis for certification and then incorporating and integrating the EMS elements of ISO 14001, we will describe the basic steps involved in the certification process for ISO 9001, inclusive of the requirements of ISO 14001. Preparation for certification requires the completion of the following items:

- Choose the level of conformance standard from ISO 9001, 9002, or 9003 for which certification is required. We have chosen ISO 9001 as the example because ISO 9001 has the highest set of requirements, and as such, ISO 9002 and ISO 9003 are subsets of ISO 9001. The quality system requirements of ISO 9001 are shown in Table 7-4.

Table 7-4. ISO 9001 (1994)—Quality System Elements

1. Management Responsibility
   - Quality policy
   - Organization
     - Responsibility and authority
     - Resources
     - Management representative
   - Management review

2. Quality System
   - General
   - Quality system procedures
   - Quality planning

3. Contract Review
   - General
   - Review
   - Amendment to a contract
   - Records

4. Design Control
   - General
   - Design and development planning
   - Organizational and technical interfaces
   - Design input
   - Design output
   - Design review
   - Design verification
   - Design validation
   - Design changes

5. Document and Data Control
   - General
   - Document and data approval and issue
   - Document and data changes

6. Purchasing
   - General
   - Evaluation of subcontractors
   - Purchasing data
   - Verification of purchased product
     - Supplier verification at subcontractor's premises
     - Customer verfication of subcontracted product

7. Control of Customer-Supplied Product

8. Product Identification and Traceability

9. Process Control

10. Inspection and Testing
    - General
    - Receiving inspection and testing
    - In-process inspection and testing
    - Final inspection and testing
    - Inspection and test records

11. Control of Inspection, Measuring, and Test Equipment
    - General
    - Control procedure

12. Inspection and Test Status

13. Control of Nonconforming Product
    - General
    - Review and disposition of nonconforming product

14. Corrective and Preventive Action
    - General
    - Corrective action
    - Preventive action

15. Handling, Storage, Packaging, Preservation, and Delivery
    - General
    - Handling
    - Storage
    - Packaging
    - Preservation
    - Delivery

16. Control of Quality Records

17. Internal Quality Audits

18. Training

19. Servicing

20. Statistical Techniques
    - Identification of Need
    - Procedures

- Implement the EMS/TQM system, inclusive of the requirements of ISO 9001 and ISO 14001. Ensure that all operating procedures are effectively implemented and maintained.

- Conduct internal audits of the system to ensure conformance to the requirements of ISO 9001 and ISO 14001.

- Select a suitable quality registrar for the third-party certification. The list of registrars operating in a country is generally available from the national standardization body of the country.

- Complete the total on-site audit and assessment process with the chosen registrar and meet all the specified requirements.

- Obtain certification status and maintain the system for conformance to surveillance audits by the registrar.

We will now expand on these for preparations for certification in the next few sections.

QUALITY SYSTEM DOCUMENTATION

For the effective functioning of a quality system, its procedures and processes must be properly documented and implemented. Normally, a quality system is documented by means of one or several tiers of documents. One such hierarchy is schematically shown in Figure 7-1.

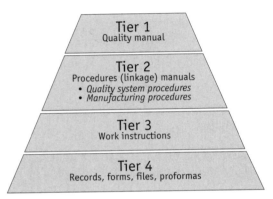

Figure 7-1. Documentation Hierarchy Pyramid

Quality Manual

The quality manual is the most important item of the system documentation. It is a silent, but powerful, spokesperson for the company. It clearly identifies to everyone the company's commitment to quality and environment through its mission and vision; quality policy; environmental policy; quality and environmental objectives and targets; and quality system and environmental procedures, processes, and methodologies.

Quality System Procedures

These procedures are an extension of the quality manual, and as such, they provide greater detail on how each system element addressed in the quality manual is implemented. Whereas the quality manual identifies the existence of a procedure, the second level system procedures are needed because the company has to identify and describe the means and methods by which the system is implemented, operated, and maintained.

Although developing this second set of procedures separately from the quality manual seems repetitive, it is necessary. Detailed operating procedures are normally confidential and internal to a company. On the other hand, a quality manual is brief and generic in nature and copies are available to anyone upon request. So if detailed system level procedures were to be combined into the quality manual, the company would run the risk of divulging their internal procedures to outside agencies.

Also, as a normal part of business and manufacturing process, operational procedures are continuously evolving and changing. If these procedures are included into the quality manual, then whenever a change was made on the procedures, it would have to be reported to the registrar as well as to anyone who has a copy of the manual. This would create an unwarranted burden on the system. Whereas, if the system level procedures are developed as a separate entity, routine changes in the procedures can be easily accommodated and managed. It must be clearly understood that significant and major operational changes, whether made in the quality manual or system level procedures, have to be reported to the registrar anyway.

Manufacturing Procedures

These are the typical Standard Operating Procedures (SOPs) through which the company runs its normal business of producing products and services. They also contain the requisite product specifications, materials, processes, and procedures. There is no specific requirement of ISO certification process for the format, structure, and extent of these SOPs; however, there is a need to interface and link these procedures with the ISO requirements as identified in the quality manual and system level procedures.

Work Instructions

Work instructions are really summaries of the SOPs. They present a brief, step-by-step set of operational instructions on how to handle a task. Many companies use these for training purposes only.

Proformas

Forms, proformas, books, records, and files make up the tangible backbone of the quality system. Some of the important ones needed for the implementation and functioning of the system are as follows:

* Document change control forms
* Nonconformance and corrective action report forms
* Audit nonconformance report forms
* Training record matrix
* Equipment calibration matrix

Quality Plans

Clause **4.2: Quality System** of ISO 9001 stresses the need and requirement for quality planning and the preparation of quality plans. The standard stipulates that the quality plans can be in the form of a reference to the appropriate documented procedures that constitute an integral part of the quality system. The ISO/TC 176 technical committee had developed a guidance standard, *ISO 10005: Quality Management—Guidelines for Quality Plans*, which can be of great assistance in developing quality plans.

Note: With respect to quality system documentation, it is important to clearly understand that ISO 9001 does not impose any restrictions on the documentation hierarchy nor any specific format or structural restrictions for the documents. It simply requires the establishment of a quality manual addressing the twenty system elements and the operational procedures (SOPs) with a linkage to the quality manual.

QUALITY MANUAL FUNDAMENTALS

The definition given for a quality manual in the international standard *ISO 8402 (1994): Quality Vocabulary* is as follows:

> **Quality Manual:** Document stating the quality policy and describing the quality system of an organization.

A quality manual describes the documented quality system procedures intended for the overall planning and administration of activities which impact the quality of an organization's products and services. Generally, the purpose of the manual is to:

- Serve as a means of communicating a company's policy, procedures, and commitment to quality

- Formalize and document the quality system

- Assist in the effective implementation and maintenance of the quality system

- Establish effective inter- and intra-organizational interfaces

- Provide an improved control of operations

- Serve as the basis for auditing quality system performance

- Provide consistency and uniformity in the application of system procedures

- Provide objective evidence that the company is truly operating a quality system as per the stated policies and objectives

Since manuals are indispensable tools and guidebooks for both inexperienced and expert workers, they must be developed with great care. They should be precise, accurate, and simple to understand. They should not be unnecessarily thick or bulky nor contain confusing or complicated instructions and procedures. Manuals should help the worker in finishing the job quickly and right the first time. If the manual is so complicated or long-winded that workers have to spend time figuring out the manual itself rather than the outlined procedures to do their jobs, then tasks will go unfinished, machines will simply not work, and the workers will become unduly demoralized.

The quality manual should be accurate, to the point, and should only contain the management mission, the quality policies, objectives, plans, and an overall description of the system elements operating in the company. Procedural details should be left for the procedural manuals and work instructions.

Similarly, the procedures manuals should also be precise and not be too complicated or bulky for the operating personnel to read and utilize. They are not meant to be glossy ornamental pieces to be put on bookshelves; they are to contain practical methods and procedures described in such a way as to let its users work with speed and accuracy to save time, effort, materials, and utilities while at the same time protecting the worker and the organization from every conceivable risk and danger. The instructions manual for a certain process or piece of equipment is no different. It should be written with clarity and precision. The operator should not be wasting time deciphering the instructions or be forced to remember unnecessary things.

By following some simple guidelines, a company can avoid creating a useless tome. For a concise, practical quality manual, a company should:

- Develop a manual that is precise, practical, and specific to the elements addressed. Even if each user has a different level of education and/or experience, the manual should be understandable by all and easy to follow.

- Develop a format and structure that is well thought out and followed consistently and uniformly throughout the manual.

- Make sure the sequence of requirements, procedures, and instructions are in line with the sequence of operations.

- Make the information precise, error free, necessary, pertinent, and directly applicable. There is no need to give long explanations and complicated theories. The illustrations, flow diagrams, or sketches should be simple and not crowded with details. The definitions, abbreviations, and terminology should be correct and uniformly acceptable to all. Remember that the user needs to read the manual, apply it, and get the job done properly and efficiently.

QUALITY MANUAL PREPARATION

When a quality manual has to be prepared to document the system elements vis-à-vis ISO 9001 and ISO 14001 certification either from existing company quality system documentation or from scratch, any or all of the recommended procedures in Table 7-5 may be used.

QUALITY MANUAL: FORMAT AND STRUCTURE

Although there is no required structure or format for the quality manual, the following guidelines should help. For the specific purpose of ISO 9001 and ISO 14001 registration, a quality manual should be prepared, chapter by chapter, responding to all the clauses outlined in ISO 9001 and ISO 14001, identifying clearly and accurately the systems and procedures operating in the company.

Typically, the manual would include, as a preamble, a series of introductory chapters (see Figure 7-2).

Preamble of the Quality Manual

Cover Page: sets out company's name and address, manual revision number, controlled copy number, person or organization to whom issued, signature and date of authority approving and reviewing the manual.

Table of Contents: provides a total list of contents addressed in the manual.

Foreword—Company Profile: presents a brief profile of the company and its operations: nature of business of the specific plant, location, or division to which the manual pertains; company's overall quality philosophy; and core environmental values and beliefs.

Introduction to Manual: identifies scope and applicability of the manual: overall framework of quality and environmental systems and responsibilities.

Table 7-5. Quality Manual Preparation Checklist For ISO 9001 and 14001 Certification

Recommended Procedures	Date Started	Date Completed	N/A
• Establish a team of competent personnel to co-ordinate and develop the manual.			
• Identify and study the applicable requirements of ISO 9001 and ISO 14001.			
• Identify the existing applicable quality and environmental system procedures operating in the company.			
• Obtain an up-to-date status of the quality and environmental systems from each function and activity in the organization.			
• Establish the format and structure of the intended manual.			
• The team, under its leader, starts the actual writing activity. The team leader may:			
– use the services of an outside consultant, if needed			
– delegate portions of the manual writing activity to other functional units, as appropriate			
– continually seek up-to-date information from various functions, as required			
• When the draft of the manual is ready, it should be reviewed by:			
– the steering committee or management to ensure accuracy of the statements relating to management commitment, and quality and environmental policies and objectives			
– the various Process Management Teams or functions of the organization to ensure completeness and accuracy of the system elements, procedures, processes, and methodologies			
– other essential personnel, as appropriate			
• The ultimate responsibility for ensuring the completeness and accuracy of the manual as well as of its contents and writing style lies with the appointed team.			
• Before issuing the manual, the document should be subjected to a final review by the appropriate responsible personnel.			
• Finally, procedures must be established for the distribution, control, development, review, and revision of the manual.			

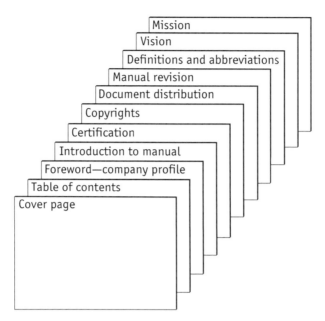

Figure 7-2. Quality Manual Preamble

Certification: certification to the effect that the quality manual adequately and accurately describes the quality and environmental systems in use within the company. This page must be signed and dated by the Chief Executive officer and the Quality Manager.

Copyrights: clearly stipulates the rights for the distribution and reproduction of the manual.

Document Distribution: identifies document distribution policy regarding controlled and uncontrolled copies, responsibility and authority for document control system, and record keeping.

Manual Revision: stipulates how often the manual will be reviewed and revised and how the revisions would be transmitted to everyone concerned.

Definitions and abbreviations: provides the relevant definitions, abbreviations, and terminology used in the manual.

Vision: sets out company's business vision, if any. (Optional—not required by ISO).

Mission: Presents company's mission statement. (Optional—not required by ISO).

Format and Structure

The format and structure of the manual also requires careful consideration. A standardized format can be developed to achieve consistency and uniformity. For instance, the top of each page in the manual should bear the company name and logo (if practical), the title of the ISO 9001 clause being addressed and the document control information, such as: page number, section, issue, date, and approving authority.

The description of the system elements for each clause can also be categorized under standardized headings (see Table 7-6), such as:

- Purpose
- Scope
- Responsibility
- General Procedures
- References

After the introductory chapters and the format and structure of the quality manual have been developed, the manual then addresses, clause by clause, all twenty chapters of the system elements of ISO 9001 as well as the ISO 14001 clauses. Each clause describes what is required to be included in the Quality Manual to meet the ISO 9001 and ISO 14001 requirements. One should follow each clause, item by item, and identify all the management systems, procedures, and processes operating in the company. To facilitate the development of the quality manual, a useful *manual checklist* is presented in the next section. The checklist is a step-by-step list of requirements for each clause of ISO 9001 with the requirements of ISO 14001 properly incorporated. By writing the quality manual with the help of this checklist, a company can be sure of adequately addressing the requirements of both standards.

Table 7-6. Quality Manual—Sample Format

LOGO ABC Company, Inc. Lovers Lane Quality Land	**Quality Manual**	**Section: 4.5** Issue: 1 Page: 1 of 3 Date: 7–27–95 Approved by:

Document and Data Control

Purpose

Scope

Responsibility

General procedures

General (4.5.1: ISO 9001)
Document and data approval and issue (4.5.2: ISO 9001)
Document and data changes (4.5.3: ISO 9001)
Document control (4.3.5: ISO 14001)

References

INTEGRATED SYSTEM CHECKLIST: ISO 9001 AND ISO 14001

The detailed and complete checklist for the combined system elements of ISO 9001 and ISO 14001 presented in this section should provide ample guidance for developing an integrated quality manual. The integration of the two standards is shown in Table 7-7. The checklist has been formulated from the ISO 9001 system elements shown in Table 7-3 and the detailed system elements and checklist for ISO 14001 outlined in the previous chapter. The format and structure of this combined checklist follows the sequence of ISO 9001 system elements and clauses. The ISO 14001 clauses have also been identified in italics wherever they are incorporated into the sequence.

The manual should be prepared chapter-by-chapter for the twenty clauses ensuring that each chapter adequately addresses the clause elements as identified in the checklist. In addition to the preparation of the quality manual, the checklist can be used for implementing and auditing the system for certification.

It should be kept in mind that the checklist only provides a sequential flow of the quality system requirements of ISO 9001 and ISO 14001. They have to be carefully and optimally harmonized and intermeshed so as to present a coherent description of the system requirements. An ideal quality manual would be one that will clearly outline the company's procedures and processes in a systematic manner while ensuring that all the requisite quality system requirements of the integrated checklist are adequately accommodated. In Table 7-8 we provide a checklist for the system elements integration of ISO 9001 and ISO 14001.

Table 7-7. System Element Integration—ISO 9001 and ISO 14001

Clause #	ISO 9001	ISO 14001
4.1	Management Responsibility	4.1 Environmental Policy; 4.2.1 Environmental Aspects; 4.2.3 Objectives/Targets; 4.3.1 Structure/Responsibility; 4.5 Management Review
4.2	Quality System	4.2.4 Environmental Management Program; 4.3.4 Environmental Documentation
4.3	Contract Review	4.2.2 Legal and Other Requirements; 4.3.6 Operational Control
4.4	Design Control	4.2.2 Legal and Other Requirements; 4.3.6 Operational Control
4.5	Document and Data control	4.3.5 Document Control
4.6	Purchasing	4.3.6 Operational Control
4.7	Control of Customer-Supplied Product	4.3.6 Operational Control
4.8	Product Identification and Traceability	4.4.1 Monitoring and Measurement
4.9	Process Control	4.3.6 Operational Control
4.10	Inspection and Testing	4.4.1 Monitoring and Measurement
4.11	Control of Inspection, Measuring, and Test Equipment	4.4.1 Monitoring and Measurement
4.12	Inspection and Test Status	
4.13	Control of Nonconforming Product	4.4.2 Nonconformance, Corrective/Preventive Action
4.14	Corrective and Preventive Action	4.3.7 Emergency Preparedness and Response; 4.4.2 Nonconformance, Corrective/Preventive Action
4.15	Handling, Storage, Packaging, Preservation, and Delivery	4.3.6 Operational Control
4.16	Control of Quality Records	4.4.3 Records
4.17	Internal Quality Audits	4.4.4 Environmental Management System Audit
4.18	Training	4.3.2 Training, Awareness, and Competence
4.19	Servicing	4.3.3 Communication; 4.3.5 Operational Control
4.20	Statistical Techniques	

Table 7-8. Integrated System Checklist: ISO 9001 and ISO 14001

Legend

☞☞☞☞☞ Adequate Coverage
☞☞☞☞ Improvement needed in system development/ implementation

☞☞ Fails to meet criteria in system development/ implementation

☞ N/A

SYSTEM ELEMENTS	☞☞☞☞	☞☞	☞☞	☞

4.1 Management Responsibility

4.1.1 Quality Policy (ISO 9001)
- Define and document quality policy and objectives.
- Identify commitment to quality.
- Policy to be consistent with the company's goals and expectations and needs of its customers.
- Policy to be understood, implemented, and maintained at all levels.

4.1 Environmental Policy (ISO 14001)
- *Define and document the environmental policy.*
- *Policy to be commensurate with the nature and magnitude of the environmental impacts of activities.*
- *Policy to identify commitment to continuous improvement.*
- *Policy to be consistent with the relevant environmental legislation and regulations.*
- *Policy to be understood, implemented, and maintained at all levels of the organization.*
- *Policy to be made publicly available.*

4.2.1 Environmental Aspects (ISO 14001)
- *Establish and maintain procedures to identify the environmental aspects of activities throughout their life cycle stages.*
- *Evaluate and document the environmental impact of these activities.*
- *Continuously review, update, and communicate information regarding the environmental aspects of activities and their impact on everyone concerned.*

SYSTEM ELEMENTS	☞☞☞☞☞	☞☞☞	☞☞	☞

4.2.3 Objectives and Targets (ISO 14001)

- *Establish environmental objectives and targets at each relevant level and function within the organization.*

- *Environmental objectives and targets are to be consistent with the environmental policy and commitment.*

- *Environmental objectives and targets to adequately address and accommodate the environmental aspects of activities, products, and services.*

- *Objectives to be continuously reviewed in relation to the legal and regulatory requirements; nature and impact of environmental aspects of activities; technological, financial, and business requirements; and views and concerns of the stakeholders.*

4.1.2 Organization (ISO 9001)

4.1.2.1 Responsibility and Authority (ISO 9001)

- Define responsibility, authority and interrelations of personnel who manage, perform, and verify work affecting quality.

- Identify personnel with designated authority and organization freedom to:

  – identify and record quality problems

  – initiate, recommend, or provide solutions

  – take preventative actions

  – verify and ensure implementation of solutions

  – control further processing until deficiency is corrected

4.1.2.2 Resources (ISO 9001)

- Identify resource requirements, provide adequate resources, and assign trained personnel for management performance of work and verification activities including internal quality audits.

Table 7-8. Integrated System Checklist: ISO 9001 and ISO 14001 *(cont'd)*

SYSTEM ELEMENTS	☞☞☞☞	☞☞☞	☞☞	☞
4.3.1 Structure and Responsibility (ISO 14001)				
• *Define, document, and assign appropriate roles, responsibilities, and authorities for all aspects of the environmental management system.*				
• *Provide adequate human, technological, and financial resources.*				
• *Appoint a management representative, with defined responsibility and authority to:*				
– *ensure that the EMS is established, implemented, and maintained in accordance with the requirements of ISO 14001*				
– *report to the management on the performance of the system for review and improvement*				
4.1.2.3 Management Representative (ISO 9001)				
• Executive management to appoint a representative with defined authority and responsibility to:				
– ensure effective implementation/maintenance of the quality system				
– report to the management on the performance of the quality system for review and improvement				
4.1.3 Management Review (ISO 9001)				
• Management reviews to be conducted at appropriately defined intervals to ensure continuing suitability and effectiveness of the quality system in satisfying ISO 9001 requirements and company's stated policy and objectives.				
• Records of reviews to be maintained.				
4.5 Management Review (ISO 14001)				
• *Management reviews of the EMS to be conducted at regular intervals to ensure its continuing suitability, adequacy, and effectiveness.*				
• *Records of reviews to be maintained.*				
• *Management reviews to initiate changes to environmental policy, objectives, and other aspects of the EMS vis-à-vis the audit results, changing and evolving technologies or any other aspect relating to continual improvement.*				

SYSTEM ELEMENTS	☞☞☞☞	☞☞☞	☞☞	☞

4.2 Quality System

4.2.1 General (ISO 9001)

- Establish and maintain a documented quality system to ensure that product conforms to specified requirements.

- Prepare a quality manual outlining the quality system procedures and structure of the quality system documentation.

4.2.2 Quality System Procedures (ISO 9001)

- Prepare documented procedures consistent with the quality policy and effectively implement the quality procedures and system.

4.2.3 Quality Planning (ISO 9001)

- Prepare suitable quality plans consistent with the activities outlined for the implementation of the quality system.

4.2.2 Legal and Other Requirements (ISO 14001)

- *Establish and maintain procedures for identifying legal and other regulatory requirements concerning the environmental aspects of activities, products, and services.*

- *Procedures must comply with legal requirements.*

4.3.4 Environmental Management System Documentation (ISO 14001)

- *Establish and maintain information, in paper or electronic form, to:*

  - *outline the core EMS elements, their requirements and interactions*

  - *provide direction to appropriate relevant documentation*

4.3.6 Operational Control (ISO 14001)

- *Establish and maintain documented procedures for controlling the environmental impact of activities.*

Table 7-8. Integrated System Checklist: ISO 9001 and ISO 14001 *(cont'd)*

SYSTEM ELEMENTS	☞☞☞☞☞	☞☞☞	☞☞	☞
4.3 Contract Review				
4.3.1 General (ISO 9001)				
• Establish and maintain procedures for contract review and coordination of review activities.				
4.3.2 Review (ISO 9001)				
• Contract review activities to include:				
– requirements defined and documented				
– conflicting requirements resolved				
– capability of supplier to meet specified requirements				
4.3.3 Amendment to Contract (ISO 9001)				
• Establish procedures for amending the contract.				
• Ensure that amendments are reported to other functions concerned.				
4.3.4 Records (ISO 9001)				
• Records of contract review activities to be maintained.				
4.2.2 Legal and Other Requirements (ISO 14001)				
• *Ensure that the contract addresses and complies with any requisite legal and regulatory requirements concerning the environmental aspects of activities, products, and services.*				
4.3.6 Operational Control (ISO 14001)				
• *Establish and maintain procedures to control the environmental aspects of activities and ensure that the suppliers and subcontractors clearly understand the organization's environmental concerns and requirements.*				

SYSTEM ELEMENTS	☞ ☞ ☞ ☞
	☞ ☞ ☞
	☞ ☞ ☞
	☞

4.4 Design Control

4.4.1 General (ISO 9001)

- Establish and maintain procedures to control and verify product designs to ensure that specified requirements are met.

4.3.6 Operational Control (ISO 14001)

- *Establish and maintain documented procedures, stipulating operating criteria, or control activities that exert environmental impacts.*

4.4.2 Design and Development Planning (ISO 9001)

- Plans to be prepared for design and development activity.
- Design and development activity to be assigned to qualified personnel equipped with adequate resources.
- Plans to describe or reference these activities and be updated as design evolves.

4.4.3 Organizational and Technical Interfaces (ISO 9001)

- Organizational and technical interfaces to be identified.
- Necessary information documented, transmitted, and regularly reviewed.

4.4.4 Design Input (ISO 9001)

- Design inputs to be identified, documented, and reviewed for adequacy.
- Incomplete, ambiguous, or conflicting requirements to be resolved with those responsible for imposing these requirements.
- Design input to accommodate any amendments or changes in the contract.

4.2.2 Legal and Other Requirements (ISO 14001)

- *Ensure that all legal and regulatory requirements concerning the environmental aspects of activities are adequately addressed in the design input process.*

Table 7-8. Integrated System Checklist: ISO 9001 and ISO 14001 *(cont'd)*

SYSTEM ELEMENTS	☞ ☞ ☞ ☞	☞ ☞ ☞	☞ ☞	☞
4.4.5 Design Output (ISO 9001)				
• Design results to be reviewed and documented at all stages of design development.				
• Review teams to include all requisite personnel associated with function concerned.				
• Records maintained of design reviews.				
4.4.6 Design Review (ISO 9001)				
• Design output to be documented and expresses in terms of requirements, calculations, and analyses that can be verified.				
• Design output to:				
– meet input requirements				
– contain and reference acceptance criteria				
– conform to regulatory requirements				
– identify characteristics crucial for safety and proper functioning of product				
• Design output to include design review procedure before release.				
4.4.7 Design Verification (ISO 9001)				
• Design verification activities to be established, planned, documented, and assigned to competent personnel.				
• Design verification to ensure output meets input requirements, using control measures such as:				
– documented design reviews				
– undertaking qualification tests and demonstrations				
– alternative calculations				
– similarity to proven design				
– reviewing design stage documents before release				
4.4.8 Design Validation (ISO 9001)				
• Procedures to be established for design validation.				
• Validation activities to be performed on the final product or at any stage as per requirements.				

SYSTEM ELEMENTS	☞☞☞☞ ☞☞☞ ☞☞ ☞			

4.4.9 Design Changes (ISO 9001)

- Establish and maintain procedures to identify, document, review, and approve changes and modifications.
- Design changes approved by authorized personnel before their utilization.

4.5 Document and Data Control

4.5.1 General (ISO 9001)

- Procedures to be established, documented and maintained, electronically or otherwise, to control all documents and data.

4.5.2 Document and Data Approval and Issue (ISO 9001)

- Documents and data relating to quality system to be reviewed and approved by authorized personnel prior to issue.
- Master list (or equivalent) to be established to identify current issues.
- Control to ensure that:

  – pertinent issues of documents are available at appropriate locations

  – obsolete documents are promptly removed from use (they can be retained for legal and/or knowledge preservation purposes as long as they are suitably identified)

4.5.3 Document and Data Changes (ISO 9001)

- Changes to documents reviewed and approved by:

  – same authority as original approval, or

  – by designated organization with access to pertinent background information

- Nature of changes to be identified, where practical, on document or attachments.

4.3.5 Document Control (ISO 14001)

- *Establish and maintain an effective document control system. Ensure that the documents are:*

Table 7-8. Integrated System Checklist: ISO 9001 and ISO 14001 *(cont'd)*

SYSTEM ELEMENTS	☞ ☞ ☞ ☞ ☞ ☞ ☞ ☞ ☞ ☞			
- available at all requisite locations				
- periodically reviewed, revised as necessary, and approved for adequacy by authorized personnel				
- promptly removed from all points of issue or use when they become obsolete—obsolete documents can be retained for legal and/or knowledge preservation purposes as long as they are suitably identified				
- legible, readily identifiable, maintained in an orderly manner, and retained for a specified period of time				
• *Establish and maintain procedures identifying appropriate responsibilities and authorities for the creation, review, and revision of the various types of environmental documents.*				
4.6 Purchasing				
4.6.1 General (ISO 9001)				
• Ensure that purchased products conform to specified requirements.				
4.6.2 Evaluation of Subcontractors (ISO 9001)				
• Select subcontractors that can meet requirements (including quality).				
• Quality records of acceptable subcontractors established and maintained.				
• Evaluation and selection of subcontractor and controls exercised on them are to be dependent on type and complexity of product and where appropriate on records of previous performance.				
4.3.6 Operational Control (ISO 14001)				
• *Establish and maintain documented procedures to ensure that the suppliers and subcontractors are thoroughly aware of the environmental concerns and requirements of the organization and they are capable of meeting those requirements.*				
4.6.3 Purchasing Data (ISO 9001)				
• Purchasing documents to clearly describe ordered product, including, where applicable:				
– type, class, style, grade, or other precise identification				

SYSTEM ELEMENTS	☞☞☞☞	☞☞☞	☞☞☞	☞

    – title, identification, and issue of specification, drawings, etc., including approval and qualification of product, procedures, processes, and personnel

    – title/number/issue of quality system standard to be applied

4.6.4 Verification of Purchased Product (ISO 9001)

- Specify verification and product release procedures in purchasing documents.
- Customer is afforded the right to verify product at subcontractor's and supplier's premises.

4.7 Control of Customer-Supplied Product

4.7 Control of Customer-Supplied Product (ISO 9001)

- Procedures for verification/storage/maintenance of such products to be established and maintained.
- Ensure recording and reporting to the customer of lost, damaged, or unsuitable product.

4.3.6 Operational control (ISO 14001)

- *Establish procedures for monitoring and control of customer-supplied products that may exert an impact on the environment.*

4.8 Product Identification and Traceability

4.8 Product Identification and Traceability (ISO 9001)

- Procedures to be established and maintained to identify product through all stages of production, where appropriate.
- When traceability is a requirement, individual product or batches will have unique identification which is recorded.

4.4.1 Monitoring and Measurement (ISO 14001)

- *Establish procedures to monitor and track the performance and conformance of operational controls with the environmental objectives and targets.*

Table 7-8. Integrated System Checklist: ISO 9001 and ISO 14001 (*cont'd*)

SYSTEM ELEMENTS	☞☞☞☞ ☞☞☞ ☞☞☞ ☞			

4.9 Process Control

4.9 Process Control (ISO 9001)

- Identify and plan the production, installation, and servicing of processes under controlled conditions that include:

    – documented procedures and/or work instructions; use of suitable production and installation equipment; suitable working environment; and compliance with standards, codes, and quality plans

    – monitoring and control of process and product characteristics

    – approval of processes and equipment

    – workmanship standards stipulated in writing or by representative samples

    – maintenance of equipment

- **Special Processes:** where full verification of the processes by subsequent inspection and testing is not possible, continuous monitoring and/or compliance of processes with documented procedures is required to ensure that specified requirements are met.

4.3.6 Operational Control (ISO 14001)

- *Identify the operations and activities that exert significant environmental impact, and establish control procedures to ensure that these activities are carried out under specified conditions. Control procedures could include:*

    – establishing and maintaining documented procedures

    – stipulating operating criteria in the procedures

    – partnering with suppliers and subcontractors to ensure that they clearly understand the organization's environmental concerns and requirements.

4.10 Inspection and Testing

4.10.1 General (ISO 9001)
- Establish and maintain documented procedures for inspection and testing in the quality plan.

- Identify how records shall be maintained.

SYSTEM ELEMENTS	☞☞☞☞	☞☞☞	☞☞	☞

4.4.1 Monitoring and Measurement (ISO 14001)

- *Establish procedures to monitor and measure key environmental characteristics throughout the life cycle stages of activities.*

4.10.2 Receiving Inspection and Testing (ISO 9001)

- Establish quality plan or procedures to verify acceptability of incoming product prior to use.

- Establish control procedures for incoming material with the subcontractor.

- If released for urgent production prior to verification, ensure that product is identified and recorded, and can be recalled or replaced if nonconforming.

4.10.3 In-Process Inspection and Testing (ISO 9001)

- In-process control procedures to be established, including:

  – inspection, test, and identification of products using a quality plan or procedures

  – in-process inspection and tests completed or reports received prior to further processing, except when released under recall procedures

4.10.4 Final Inspection and Testing (ISO 9001)

- All final inspections and tests can be carried out as per quality plan or procedures.

- Ensure that all required receiving/in-process inspections and tests have been carried out.

- Product should not be shipped until inspection results are complete and product indicates conformance to specified requirements.

- Ensure that all inspections and tests have been completed and associated data and documentation are available and authorized prior to release.

4.10.5 Inspection and Test Records (ISO 9001)

- Records of inspections are to be maintained to identify that product has passed inspection and/or test with defined acceptance criteria.

Table 7-8. Integrated System Checklist: ISO 9001 and ISO 14001 *(cont'd)*

SYSTEM ELEMENTS	☞ ☞ ☞ ☞			

• Records must identify inspection authority responsible for release for product.

4.11 Control of Inspection, Measuring and Test Equipment

4.11.1 General (ISO 9001)
• Control, calibrate, and maintain inspection, measuring, and test equipment to demonstrate product conformance to specified requirements.

• Equipment must be used to ensure that measurement uncertainty is known and is consistent with required capability.

• Control the use, calibration, and inspection of test hardware or software when used for acceptance.

• Calibration procedures, schedule of checks/rechecks, and pertinent records should be established and maintained for test hardware and software.

• Technical data pertaining to measurement devices should be made available to customer for verification, when required.

4.4.1 Monitoring and Measurement (ISO 14001)

• *Establish procedures for calibrating the monitoring equipment used for environmental processes.*

• *Maintain records of calibration.*

4.11.2 Control Procedures (ISO 9001)
• Control procedures should include:

– identification of measurements to be made, required accuracy and selection of appropriate equipment

– calibration at prescribed intervals, or prior to use, against certified equipment traceable to nationally and internationally recognized standards, or documenting the basis used for calibration if no such standards exist

– establishment and maintenance of calibration procedures, including details of equipment type, identification number, location, frequency, check method, acceptance criteria, and action required when results are unsatisfactory

SYSTEM ELEMENTS	☞ ☞ ☞ ☞	☞ ☞ ☞	☞ ☞	☞

      – identification of equipment calibration status with indicator or record

      – maintenance of calibration records

      – assessment and documentation of the validity of previous inspection and test results when equipment is found to be out of calibration

      – suitable environmental conditions for calibration

      – handling, storage, and preservation of equipment to ensure its accuracy and fitness

      – safeguarding equipment from unauthorized adjustments/tampering

4.12 Inspection and Test Status

4.12 Inspection and Test Status (ISO 9001)

- Establish procedures for the identification of inspection and test status of product to indicate conformance or nonconformance.

- Control and maintenance of inspection and test status required, as necessary, throughout production, installation, and servicing to ensure only acceptable product is shipped, used, or installed.

4.13 Control of Nonconforming Product

4.13.1 General (ISO 9001)

- Establish and maintain documented procedures to ensure that nonconforming product is prevented from unintended use or installation.

- Control the identification, documentation, evaluation, segregation, disposition, and notification of nonconforming product to the functions concerned.

4.13.2 Review and Disposition of Nonconforming Product (ISO 9001)

- Define responsibility for review and approval of dispositions.

- Procedures to control disposition of nonconforming product may include:

      – rework

Table 7-8. Integrated System Checklist: ISO 9001 and ISO 14001 (*cont'd*)

SYSTEM ELEMENTS	☞ ☞ ☞ ☞ ☞ ☞ ☞ ☞ ☞ ☞ ☞			
– acceptance with or without repair				
– regrading				
– rejecting or scrapping				
• When required by contract, the use or repair of non-conforming product should be reported to customer for concession.				
• Record actual condition of accepted or repaired non-conforming product.				
• Establish and document plans and procedures for reinspection of repaired or reworked product.				
4.4.2 Nonconformance and Corrective/Preventive Action *(ISO 14001)*				
• *Establish procedures for defining responsibility and authority for handling nonconformance of the environmental issues and for taking corrective action.*				
4.14 Corrective and Preventive Action				
4.14.1 General (ISO 9001)				
• Establish and maintain documented procedures for corrective and preventive action.				
• Corrective and preventive action to be commensurate with the magnitude/risk of problem.				
• Changes resulting from corrective and preventive action to be implemented and recorded.				
4.14.2 Corrective Action (ISO 9001)				
• Corrective action procedures should include:				
– the effective handling of customer complaints and reports of product nonconformities				
– investigating the cause of nonconformities relating to product, process, and quality system and recording the results of the investigation				
– determining the corrective action needed to eliminate the cause of nonconformities				
– applying controls to ensure that corrective action is taken and that it is effective				

SYSTEM ELEMENTS	☞☞☞☞	☞☞☞	☞☞	☞

4.14.3 Preventive Action (ISO 9001)

- Preventive action procedures should include:

  – the use of appropriate sources of information such as processes and work operations which affect product quality, concessions, audit results, quality records, service reports, and customer complaints to detect, analyze, and eliminate potential causes of nonconformities

  – determining the steps needed to deal with any problems requiring preventive action

  – initiating preventive action and applying controls to ensure that it is effective

  – ensuring that relevant information on actions taken, including changes to procedures, is submitted for management review

4.3.7 Emergency Preparedness and Response (ISO 14001)

- *Establish and maintain procedures to respond to environmental emergencies.*

- *Continually review and revise the emergency response procedures.*

- *Periodically test procedures, where applicable.*

4.4.2 Nonconformances and Corrective/Preventive Action (ISO 14001)

- *Establish procedures for defining responsibility and authority for handling nonconformances of environmental issues and for taking corrective and preventive action.*

- *The corrective and preventive action should be commensurate with the magnitude of the risk encountered through the environmental impact.*

- *Implement and record any changes in the documented procedures resulting from the corrective and preventive actions.*

4.15 Handling, Storage, Packaging, Preservation and Delivery

4.15.1 General (ISO 9001)

Table 7-8. Integrated System Checklist: ISO 9001 and ISO 14001 *(cont'd)*

SYSTEM ELEMENTS	☞☞☞☞		
• Establish and maintain documented procedures for handling storage, packaging, preservation, and delivery.			
4.3.6 Operational Control (ISO 14001)			
• *Establish suitable handling, storage, packaging, preservation, and delivery procedures for all those activities which exert impact on the environment.*			
4.15.2 Handling (ISO 9001)			
• Provide methods and means to prevent damage and deterioration.			
4.15.3 Storage (ISO 9001)			
• Control storage conditions to prevent damage and deterioration.			
• Control issue and receipt of goods to and from secure storage areas.			
• Control and assess the condition of stored product at appropriate intervals.			
4.15.4 Packaging (ISO 9001)			
• Control packing, packaging, and marking of product to ensure conformance to specified requirements.			
4.15.5 Preservation (ISO 9001)			
• Establish methods for preservation and segregation of product from the time of receipt until the supplier's responsibility ceases.			
4.15.6 Delivery (ISO 9001)			
• Control protection of product quality after final inspection, test, and acceptance.			
• When contractually specified, protection to include delivery to destination.			
4.16 Control of Quality Records			
4.16 General (ISO 9001)			

SYSTEM ELEMENTS	☞☞☞☞ ☞☞☞ ☞☞☞ ☞☞☞ ☞			

- Establish and maintain documented control procedures for identification, collection, indexing, access, filing, storage, maintenance, and disposition of quality records.
- Records must demonstrate achievement of
  - required quality
  - effective operation of quality system
- Records shall include pertinent quality records of subcontractors.
- Records are to be legible and identifiable to the product.
- Records should be readily retrievable and stored in suitable environment to minimize damage or deterioration and prevent loss.
- Retention time of quality records to be established and documented.
- Where agreed contractually, records shall be made available to the customer for evaluation.

4.4.3 Records (ISO 14001)

- *Establish and maintain procedures for the identification, maintenance, and disposition of environmental records.*
- *Maintain training records as well as records of the results of audits and reviews.*
- *Records should be legible, identifiable, and traceable to the activity.*
- *Records should be stored and maintained in such a way that they are easily retrievable and protected against damage, deterioration, or loss.*
- *Retention time of records should be established and recorded.*
- *Records should be maintained to demonstrate conformance to the requirements of ISO 14001.*

4.17 Internal Quality Audits

4.17 Internal Quality Audits (ISO 9001)

Table 7-8. Integrated System Checklist: ISO 9001 and ISO 14001 *(cont'd)*

SYSTEM ELEMENTS	☞☞☞☞	☞☞☞	☞☞	☞

• Documented procedures to be established to conduct and control internal quality audits ensuring:
  – compliance with planned activities
  – effectiveness of the quality system
• Audits to be scheduled on the basis of status and importance of activity and should be conducted by staff independent of the function being audited.
• Results of the audit to be documented and reviewed by management of audited areas.
• Corrective action to be taken on the deficiencies, and its implementation to be verified.
 	
4.4.4 Environmental Management System Audit (ISO 14001)
• *Establish and maintain a program and procedures for carrying out regular audits of the EMS.*
• *The audits should determine whether the EMS has been properly implemented and maintained and conforms to the planned arrangements as per the requirements of ISO 14001.*
• *Results of the audit should be reported to the management for review and improvement of the system.*
• *The frequency and intensity of audits should be commensurate with the environmental importance of the activity concerned.*
 	
4.18 Training
 	
4.18 Training (ISO 9001)
• Establish and maintain procedures to identify training needs and provide training to those personnel performing activities affecting quality.
• Personnel performing specific tasks must be qualified on the basis of education, training, and/or experience.
• Training records are to be maintained.
 	
4.3.2 Training, Awareness, and Competence (ISO 14001)
• *Identify training needs and provide appropriate training to all personnel, commensurate with the nature of their activity and the significance of its impact on the environment.*				

SYSTEM ELEMENTS	☞☞☞☞	☞☞☞	☞☞☞	☞

- *Establish procedures to ensure that employees at all levels of the organization are aware of:*

  - *the importance of conformance to environmental procedures, policy, and EMS requirements*

  - *the environmental benefits that would accrue from improved personal performance*

  - *their role and responsibilities in achieving conformance to the environmental policy, procedures, EMS requirements, and the emergency response requirements.*

  - *the potential consequences of departure from the specified operating procedures*

- *Personnel whose work involves activities which may have a significant impact on the environment should be selected on the basis of appropriate education, training and experience.*

4.19 Servicing

4.19 Servicing (ISO 9001)

- When specified by contract, documented procedures shall be established and maintained for performing and verifying service activities.

4.3.3 Communication (ISO 14001)

- *Establish a servicing procedure to ensure that the external stakeholders are properly kept informed of the environmental aspects of activities, products, and services and the effectiveness of the organization's environmental management system to control these activities.*

4.20 Statistical Techniques

4.20.1 Identification of Need (ISO 9001)

- Establish control procedures for the application and use of statistical techniques for verifying the acceptability of process capability and product characteristics.

- Identify need for statistical techniques required to establish, control, and verify process capability and product characteristics.

4.20.2 Procedure (ISO 9001)

- Establish procedures to implement and control application of appropriate statistical techniques.

INTEGRATED SYSTEMS: IMPLEMENTATION ROAD MAP

Table 7-9 summarizes our discussion on the subject of integrated EMS/TQM system implementation and certification, and outlines a simple ten-phase generic road map for the process.

Table 7-9. Road Map For EMS/TQM Implementation: ISO 9001 and ISO 14001 Certification

Phase 1: Management Readiness

- Executive/senior management orientation and awareness
- Top management approval, support, and commitment
- Establish a steering committee
- Identify EMS/TQM/ISO coordinator/project leader

Phase 2: Strategic Planning: Needs Analysis

- Identify total requirements of ISO 9001 and ISO 14001
- Develop a master plan for implementation
- Formulate time schedule
- Identify expenditure of resources
- Identify need for external consultant

Phase 3: Current System: Evaluation

- Identify current systems, procedures, and processes
- Identify current documentation: quality manual, procedures manuals, work instructions
- Outline current infrastructure/responsibilities
- Perform a gap-analysis to identify deficiencies in the current system relative to the requirements of ISO 9001 and 14001

Phase 4: Implementation Framework

- Delineate responsibilities/authorities
- Establish Process Management Teams
- Organize a team for developing system documentation: quality manual, quality system procedures
- Identify coordinators and teams for developing SOPs, procedures, work instructions

Phase 5: ISO 9000/ISO 14001 Training

- Management training
- Team leaders and team members training
- General ISO 9000 and ISO 14000 awareness training for everyone
- Internal lead auditor/auditor training

Phase 6: Documentation Preparation

- Steering Committee to develop:
  - Mission and vision statement
  - Quality and environmental policy, objectives, plans
- Development of:
  - quality manual
  - quality system procedures
- Development and/or augmentation of:
  - manufacturing procedures
  - environmental procedures
  - standard operating procedures
  - work instructions

Phase 7: System Implementation

- Implement all requisite quality system and environmental procedures
- Integrate and interface the system
- Document the system implementation activities
- Eliminate system deficiencies, if any

Phase 8: System Conformance Audits

- Develop internal audit schedule
- Conduct section-by-section audits
- Conduct a full system audit
- Take corrective and preventive action on deficiencies
- Document the audit findings
- Execute management review and evaluation of the system

Phase 9: Registrar Certification

- Establish registration protocol with the chosen registrar
- Submit quality manual and/or system level procedures for assessment

Table 7-9. Road Map For EMS/TQM Implementation ISO 9001 and ISO 14001 Certification *(cont'd)*

- Comply with nonconformances identified by the registrar, if any
- Undergo an on-site pre-registration audit by the registrar (optional)
- Complete the comprehensive on-site certification audit by the registrar
- Take requisite action to eliminate deficiencies identified by the registrar
- Obtain certification status

Phase 10: System Maintenance and Continuous Improvement

- Conduct frequent checks and audits
- Maintain and control procedures, systems, and documentation
- Maintain compliance with surveillance audits by the registrar
- Maintain focus on continuous improvement

8

ENVIRONMENTAL QUALITY AUDITING

Auditing is a subject of profound importance for most business activities and is a valuable instrument to verify and improve the effectiveness of any system. The scope of auditing spans numerous types of situations; for example, quality auditing, environmental auditing, financial and revenue auditing, management auditing, and legal auditing. In this chapter, we shall consider the process of auditing as it is applicable to environmental quality management systems and total quality management systems for products and services. However, since auditing is a generic subject, its basic principles and framework are universally applicable.

A quality audit provides objective evidence of the effectiveness of the quality system for products, services, and environmental activities. It furnishes a timely comprehensive status report on the health of a company's quality system. Typically, a quality audit is a verification tool that determines system weaknesses and potential problems and, in so doing, provides avenues for corrective action and system improvement. The wide variety of benefits of using a quality system audit are shown in Figure 8-1.

AUDITING STANDARDS

Quality system auditing is an essential and integral part of the certification process to ISO 9000 quality system standards or ISO 14000 environmental management system standards.

The certification process involves two types of audits:

Figure 8-1. Benefits of a Quality Audit

- Internal quality audits

- External quality audits

Internal quality audits are performed on the company's own functions by the company's own personnel. They constitute an essential requirement for ISO 9000 and/or ISO 14000 certification. Companies seeking certification have to develop their own strategic internal quality audit plans and procedures.

External quality audits are performed either on the suppliers to assess their capabilities in meeting specified requirements or on the customers to assess their needs and expectations. As indicated earlier, for a third-party certification to ISO 9000 and/or ISO 14000 standards, the external quality audits are performed by the accredited registrars to assess the capabilities of the organization to meet the requirements of the stipulated standards.

Since the auditing function is an essential element of the quality and environmental management systems, both technical committees of the ISO, ISO/TC 176 on *Quality Management and Quality Assurance* and ISO/TC 207 on *Environmental Management Systems*, have developed standards on auditing to help establish effective auditing programs as well as to use for auditing the systems for certification.

Auditing standards developed by ISO/TC 176, though numbered under the ISO 10000 series, are an integral part of the ISO 9000 series of quality system standards. Thus far, the following standards have been developed and used:

ISO 10011: Guidelines for Auditing Quality Systems

Part 1: Auditing

Part 2: Qualification Criteria for Quality System Auditors

Part 3: Management of Audit Programs

Environmental auditing standards are still in the process of being developed by ISO/TC 207. As of this writing, three committee drafts (CD) of international standards have been prepared:

ISO/CD 14010:   Guidelines for Environmental Auditing—General Principles of Environmental Auditing

ISO/CD 14011/1: Guidelines for Environmental Auditing—Audit Procedures— Part 1: Auditing of Environmental Management Systems

ISO/CD 14012:   Guidelines for Environmental Auditing—Qualification Criteria for Environmental Auditors

In the following sections, we will present the basic generic framework of auditing practices and principles, and outline their specific applications as stipulated in the ISO 10000 and ISO 14000 series of standards. The reader is advised to consult the ISO 14000 Environmental Auditing standards for more details.

AUDITING FUNDAMENTALS

The definition of the terms *quality audit* and *environmental audit*, as given in the respective ISO 10011/1 and ISO 14011/1 standards, are as follows:

> ***ISO 10011/1: Quality Audit:*** A systematic and independent examination to determine whether quality activities and related results comply with planned arrangements and whether these arrangements are implemented effectively and are suitable to achieve objectives.

> ***ISO 14011/1: Environmental Management Systems Audit:*** A systematic and documented verification process of objectively obtaining and evaluating evidence to determine:

> • Whether an organization's environmental management system conforms to the EMS audit criteria

> • Whether the system is implemented effectively to achieve the organization's policies and objectives

> • Whether the system calls for communicating the results to the client

Audits are normally carried out for such purposes as evaluating the effectiveness of the implemented quality system; assessing conformance to specified requirements; identifying shortcomings in the system; and identifying improvement opportunities.

Quality and environmental system audits, whether internal or external, are performed with respect to a specific purpose and in accordance with a specified plan, procedure, and criteria. There are different types of quality audits and, depending upon the need or the situation, one or several of them can be performed simultaneously. Following are the basic types of quality audits:

System Audit: assessing how the quality and environmental management procedures are applied in practice and how effective they are.

Product/Service Audit: evaluating the conformity of the environmental activities, products and services to specified technical requirements.

Operational Audit: assessing the performance of a supplier/internal department or the needs and expectations of a customer.

Process Audit: verifying how closely the established methods/procedures are followed in actual practice.

Monitoring Audit: verifying the processes to confirm that all parameters are maintained within their specifications.

DEVELOPING INTERNAL AUDIT SYSTEM

One of the basic requirements for any company seeking ISO 9000 and/or ISO 14000 certification is the establishment and implementation of a comprehensive quality audit system to verify the conformance and effectiveness of the planned quality and environmental activities. Table 8-1 lists required steps.

Table 8-1. Developing a Quality Audit System—Basic Action Steps

Basic Steps To Be Taken	Date started	Date completed	Status	N/A
Define the purpose and scope of the audit.				
Establish goals and objectives.				
Identify a management commitment and focus.				
Appoint a lead auditor with designated responsibility and authority to take action.				
Establish an audit team.				
Establish an overall planning framework for the audit system.				
Define the parameters and boundaries of each activity to be audited.				
Develop implementation plans.				
Develop and document audit schedule, plans, procedures, and instructions.				
Identify resources and personnel.				
Establish priorities and action plans.				
Document the audit findings.				
Bring the audit results to the attention of the personnel having the responsibility in the audited area.				

Basic Steps To Be Taken	Date started	Date completed	Status	N/A
Take corrective and preventive action on the deficiencies identified by the audit.				
Assess the correctiveness of corrective action.				
Assess conformance to the specified requirements.				
Assess the effectiveness of the quality and environmental systems.				
Identify opportunities and initiatives for improvement.				

The success of an audit program is, typically, dependent on the following:

- **A comprehensive audit plan:** a total understanding of the quality and environmental system requirements and team effort.

- **A detailed documented set of procedures and instructions:** everyone must know, understand, and follow uniform procedures.

- **Qualified and objective auditors:** extensive audit training.

- **Thorough and unbiased reports:** qualified personnel, commitment, training, and independence of operability.

- **Documentation and communication:** an effective documentation system and reporting of deficiencies within and across all activities.

- **Timely and effective corrective action:** management commitment, resources, authority, and total cooperation.

- **System elements checklist:** based on a clear understanding of the quality system requirements of ISO 9001 and ISO 14001, it ensures that everything that needs to be done has been done properly.

QUALITY AUDIT FRAMEWORK

A quality and environmental audit may be required in a variety of situations, such as a process audit, supplier audit, audit of the environmental quality system within a company, audit by an external auditor vis-à-vis accreditation requirements, or at the request of company's management and/or a customer. In each case, the fundamental requirement is the establishment of a total audit procedural framework. Every audit procedure, whether for an internal audit or external audit, has a set of elements which must be addressed to develop the audit framework with an implementation plan. The following description of the environmental quality audit system elements should provide sufficient assistance in developing an effective audit program. Figure 8-2 schematically presents these system elements.

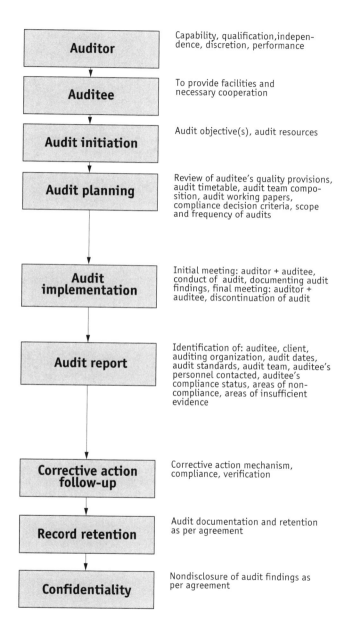

Auditor	Capability, qualification, independence, discretion, performance
Auditee	To provide facilities and necessary cooperation
Audit initiation	Audit objective(s), audit resources
Audit planning	Review of auditee's quality provisions, audit timetable, audit team composition, audit working papers, compliance decision criteria, scope and frequency of audits
Audit implementation	Initial meeting: auditor + auditee, conduct of audit, documenting audit findings, final meeting: auditor + auditee, discontinuation of audit
Audit report	Identification of: auditee, client, auditing organization, audit dates, audit standards, audit team, auditee's personnel contacted, auditee's compliance status, areas of non-compliance, areas of insufficient evidence
Corrective action follow-up	Corrective action mechanism, compliance, verification
Record retention	Audit documentation and retention as per agreement
Confidentiality	Nondisclosure of audit findings as per agreement

Figure 8-2. Quality and Environmental Audit System Elements

Auditor

The effectiveness of the audit procedures and the confidence placed in them are highly dependent on the auditor's ability and expertise. The auditor must be well qualified, professionally proficient, and adequately experienced in the subject. For the specific task at hand, the auditor's qualifications must be mutually acceptable to both the client requesting the audit and the auditing organization. Following are some of the general attributes that should be kept in mind in the selection of auditors:

- Competence in interpersonal and communication skills

- Ability to plan, organize, initiate, control, and analyze

- Leadership abilities—to supervise, delegate, support, and give direction

- Ability to work systematically, independently, and judiciously

- Ability to use discretion regarding the confidentiality and proprietary of information and audit findings

- Balanced personality; absolute honesty and integrity; good attitude, conduct and appearance; self-confidence

- Ability to exercise independence of judgment

Auditee

It is the auditee's responsibility to ensure that the audit team has been provided with adequate working facilities, access to relevant information, and effective cooperation in all matters relating to the audit.

Audit Initiation

Before initiating the audit, the audit objectives must be set and agreed upon by the parties. Basically, the audit objectives should center around verifying conformance to the stated policy, objectives, goals, system procedures, and processes. The auditing organization must clearly identify the requisite audit resources commensurate with the task at hand, and plan all the other administrative details necessary to carry out the audit effectively.

Audit Planning

Typically, audit planning consists of the following activities:

- Review of auditee's provisions: quality and environmental system documentation, specifications, standards, etc.

- Audit timetable: audit dates, schedule of meetings, audit team composition and structure, activities to be audited, and the requisite standard against which the audit is to be conducted

- Audit working papers: all relevant forms, records, checklists, assignment sheets, agenda items, etc. necessary for the conduct of the audit

- Compliance decision criteria: all the relevant criteria used in making a decision on the conformance/nonconformance of the quality and environmental system elements to the stipulated standard

Audit Implementation

Audit implementation begins with an initial meeting between the auditor and auditee management to clarify the overall plan for the conduct of the audit. Then a detailed procedure for the conduct of the audit is developed as well as a mechanism for documenting and recording the audit findings. Finally, a meeting between the parties is arranged to discuss the nature and extent of noncompliance; how it shall be documented and reported; and a mutual agreement on the discontinuation of the audit.

Audit Report

The audit report is the document that formally communicates the findings of the audit to the client and the auditee. It must be prepared with great care and tenacity. It must be thorough and detailed and would normally identify the following items: the auditee, client and auditing organization, the audit dates, the audit standards, the audit team, the auditee's personnel contacted during the audit, the auditee's compliance status, the areas of noncompliance or of insufficient evidence.

Corrective Action Follow-up

The objective here is to ensure that effective corrective action has been taken in every area of noncompliance. A follow-up procedure must be put in place to verify the effectiveness of the corrective action. Experience has indicated that companies who are able to develop self-correcting systems can effectively establish a constancy of compliance and improvement.

Record Retention

The objective here is to ensure that procedures have been clearly identified for the retention of audit records and documentation.

Confidentiality

Confidentiality is fundamental to all audit procedures. Consequently, a clear written agreement must be reached between the parties regarding the disclosure/nondisclosure of audit findings.

ISO 14011: AUDIT PROCEDURES

Outlined in Table 8-2 are the salient features of the international standard, *ISO 14011/1: Guidelines for Environmental Auditing—Audit Procedures—Part 1: Auditing of Environmental Management Systems*. The standard provides audit procedures for the planning and performance of an environmental audit for the EMS to verify its conformance to ISO 14001.

The overall framework and principles outlined in ISO 14011/1 are based on the general quality audit procedures and principles stipulated in ISO 10011/1.

Table 8-2. ISO 14011/1 Audit Procedures

System Elements	Description
4. Environmental Management System Audit Objectives, Roles, and Responsibilities	
4.1 Audit Objectives	• Define and document EMS audit objectives. • The audit objectives may include: – assessing an organization's EMS against ISO 14001 – assessing the suitability and effectiveness of EMS implementation – helping the organization in improving their EMS – carrying out an audit assessment of the EMS for possible contractual purposes
4.2 Roles, Responsibilities, and Activities	
4.2.1 Auditors	• Appoint a lead auditor to oversee the audit function.
4.2.1.1 Audit Team	• The audit team members should be selected on the basis of their qualifications, skills and experience; and also on the basis of the nature and type of functions being audited.
4.2.1.2 Auditor	The auditor's responsibilities should include: • Planning and implementing the audit effectively • Collecting and analyzing adequate information to draw objective conclusions about the performance of the EMS • Documenting individual findings • Safeguarding audit documents • Preparing working documents under the direction of the lead auditor • Supporting and assisting the lead auditor • Verifying the effectiveness of the corrective actions, if required

System Elements	Description
	• Meeting the requirements of ISO 14011/1
4.2.1.3 Lead Auditor	The lead auditor is responsible for: • All phases of the audit and the associated decisions regarding the conduct of the audit • Obtaining appropriate information to meet audit objectives • Reviewing and assessing EMS documentation against EMS audit criteria • Forming the audit team • Identifying and resolving any disagreements between the auditors and auditee • Defining objectives and requirements of each audit assignment • Developing audit plan, procedures, and schedule, and communicating to the client, auditors, and auditee • Coordinating the audit proceedings • Resolving any problems arising during the audit • Reporting audit findings to the auditee and client • Meeting the requirements of ISO 14011/1
4.2.2 Client	It is the responsibility of the client to: • Determine the need for the audit, establish its scope and objectives, and initiate the process • Select lead auditor • Provide appropriate authority and resources for the audit • Receive the audit report and initiate any follow-up action, if required
4.2.3 Auditee's Assignment	It is the responsibility of the auditee's management to: • Inform the relevant employees about the objectives and scope of the audit • Provide adequate resources for the

Table 8-2. ISO 14001/1 Audit Procedures (cont'd)

System Elements	Description
	audit • Designate responsible personnel to assist the audit team • Provide access to facilities and materials requested by the auditors • Provide full cooperation to the auditors
5. Auditing	
5.1 Initiating the Audit	
5.1.1 Audit Scope	• Define the scope of the audit commensurate with the EMS audit criteria and the physical location and nature of activities of the organization. • Scope to be developed by input from the client, lead auditor, and the auditee.
5.1.2 Preliminary Review of Auditee's Environmental Management System	• Auditor should review all pertinent documentation and information pertaining to auditee's environmental management system as a basis for planning the audit.
5.2 Preparing the Audit	
5.2.1 Audit Plan	Prepare an audit plan to include, if applicable: • Audit location and schedule • Audit objective and scope • Audit methodology • Identification of activities with significant environmental impact • Identification of audit team members • Reference documents • Unit/function to be audited • Confidentiality requirements • Format, structure and date of issue of the audit report • Document retention requirements • Reporting procedure of audit plan and findings to the client and the

System Elements	Description
	auditee
	• Procedures for the resolution of any objections or discrepancies
5.2.2 Audit Team Assignments	• Assign each auditor specific aspects and elements of the EMS, functions, or activities to audit.
	• Establish specific methodology to follow.
	• Lead auditor is to be responsible for the overall audit proceedings or for making any changes in the audit procedures during the audit.
5.2.3 Working Documents	Working documents required to facilitate and support the auditor's investigation should include:
	• Requisite forms for recording the findings
	• Checklists and methodologies for the audit
	• Records of meetings
	• Forms for reporting audit findings
5.3 **Executing the Audit**	
5.3.1 Opening Meeting	Arrange an opening meeting to:
	• Introduce the audit team
	• Review the audit plan, scope, and objectives
	• Provide a short summary of audit procedure to the auditee
	• Establish formal communication between the audit team and the auditee
	• Agree on audit schedule
	• Confirm the availability of adequate resources and facilities to conduct the audit
	• Promote active participation by the auditee
5.3.2 Collecting Evidence	• Auditors should collect all relevant and pertinent evidence required to make objective decisions regarding the

Table 8-2. ISO 14001/1 Audit Procedures (cont'd)

System Elements	Description
	effectiveness of the environmental management system.
5.3.3 Audit Findings	• Document all audit findings. • Nonconformance should be clearly identified. • Audit findings should be reviewed by the lead auditor along with the responsible auditee manager.
5.3.4 Closing Meeting with the Auditee	• Present a description and a preliminary report of the audit findings to the auditee at the closing meeting. • Discrepancies, if any, about the audit findings should be resolved with the auditee before the issuance of the final audit report.
5.4 Audit Report and Records	
5.4.1 Audit Report Preparation	• Audit report should be prepared under the direction of the lead auditor. • The lead auditor is responsible for the accuracy and completeness of audit report.

5.4.2 Report Content	• The audit report should be formal in style, format, and content.
	• The audit report should include: Identification of auditee's and client's organization
	  – auditee's representatives
	  – audit plan
	  – audit team
	  – audit objective and scope
	  – audit criteria
	  – audit process
	  – audit conclusions about the effectiveness of the EMS
	  – distribution list for the audit report
	  – statement on the confidentiality of the audit findings

5.4.3 Report Distribution	• The distribution of the audit report

System Elements	Description
	should be according to the client's requirements and instructions.
5.4.4 Record Retention	• Audit documents should be retained for the period agreed upon by the client, auditee, and the lead auditor.
6. Audit Completion	• The audit proceedings are completed once all activities relating to the agreement between the client, the auditee, and the lead auditor have been completed.
7. Corrective Action	• It is the responsibility of the auditee to take appropriate action vis–à–vis the audit findings.

ISO 10011: QUALITY SYSTEM AUDITING

Presented in Table 8–3 is an overall format of ISO 10011/1 to elucidate a comparative reference to ISO 14011/1 and to highlight the similarities between them. For more details, the reader is advised to consult the source document.

Table 8-3. ISO 10011/1—Audit Procedures

System Elements	Description
4. **Audit Objectives and Responsibilities** **4.1** **Audit Objectives**	• Define and document audit objectives. • For one or more of the following purposes audits are conducted to: – determine conformity of quality system elements with specified requirements – assess the effectiveness of system implementation – assess compliance with regulatory requirements – help improve the quality system – evaluate the quality system against a quality system standard for certification – evaluate the quality system against customer's requirements
4.2 **Roles and Responsibilities** 4.2.1 Auditors 4.2.1.1 Audit Team	• Appoint a lead auditor to oversee the audit function. • Depending on the nature and circumstances of the audit, the audit team may include observers, audit trainees, technical persons, etc.
4.2.1.2 Auditor's Responsibilities	Auditor's responsibilities include: • Complying with audit plan and requirements • Communicating audit requirements • Carrying out the planned and assigned responsibilities • Documenting and reporting the audit findings • Verifying the effectiveness of the corrective actions taken as a result of the audit • Retaining and safeguarding documents and reports pertaining to the audit

System Elements	Description
	• Supporting the lead auditor
4.2.1.3 Lead Auditor's Responsibilities	The lead auditor is responsible for: • All phases of the audit and the associated decisions regarding the conduct of the audit • Assisting in the make–up of audit team • Preparing the audit plan • Ensuring that the audit has been conducted as per the planned arrangements • Submitting the audit report • Forwarding the audit report to the management for review and action, if required
4.2.1.4 Independence of the Auditor	• Auditors must be objective and free from any influence or bias • Persons and units involved with the audit should support the independence and integrity of the auditors
4.2.1.5 Auditors' Activities	The lead auditor will: • Define requirements for each audit • Establish auditor qualifications and assignments • Comply with applicable audit requirements and appropriate directives • Plan the audit and prepare working documents • Brief the audit team • Review quality system documentation for adequacy • Report critical nonconformances to the auditee • Report audit findings clearly It is the auditor's responsibility to: • Work within the framework of audit scope • Exercise independence and objectivity • Collect and analyze evidence and draw objective conclusions • Report the audit findings

Table 8-3. ISO 10011/1—Audit Procedures (cont'd)

System Elements	Description
	• Act in an ethical manner at all times
4.2.2 Client	It is the responsibility of the client to: • Determine the need, purpose, and scope of the audit • Initiate the process • Determine the auditing organization • Assess the audit report to determine what follow–up action needs to be taken with the auditee
4.2.3 Auditee	It is the responsibility of the auditee's management to: • Inform the relevant employees about the objectives and scope of the audit • Provide adequate resources for the audit • Designate responsible personnel to assist the audit team • Provide access to facilities and materials requested by the auditors • Provide full cooperation to the auditors • Determine and initiate corrective actions based on the audit report
5. Auditing **5.1 Initiating the Audit** 5.1.1 Audit Scope	• Define scope of the audit in terms of the quality system elements, physical location, and organizational activities. • Scope to be developed by input from the client, lead auditor, and the auditee. • Scope and depth of the audit to be commensurate with the specific needs of the client. • Sufficient objective evidence should be available to demonstrate the effectiveness of the quality system. • Adequate resources should be committed to the audit commensurate with the scope and depth of the audit.

System Elements	Description
5.1.2 Audit Frequency	• Audit frequency is determined by the need and requirements of the client, the regulatory requirements, and the nature of changes that may affect the quality system. • Frequency of internal audits is determined by the management in light of their business plan and/or system effectiveness concerns.
5.1.3 Preliminary Review of Auditee's Quality System Description	• Auditor should review all pertinent documentation and information pertaining to auditee's quality system as a basis for planning the audit.
5.2 Preparing the Audit	
5.2.1 Audit Plan	• Audit plan to be approved by the client and communicated to the auditors and auditee. The plan should include: – audit scope and objectives – identification of individuals bearing responsibility for the objectives and scope – identification of quality system and reference documents – audit team – language of the audit – audit location, schedule – unit/function to be audited – confidentiality requirements – format, structure, and date of issue of the audit report
5.2.2 Audit Team Assignments	• Lead auditor to assign specific quality system elements or functional departments to audit, after consultation with the auditors concerned.
5.2.3 Working Documents	• Working documents required to facilitate and support the auditor's investigation should include: – checklist for evaluating the quality system elements – forms for reporting audit findings

Table 8-3. ISO 10011/1—Audit Procedures (cont'd)

System Elements	Description
	– forms for documenting supporting evidence for the audit findings • Working documents should be flexible enough to accommodate any additional audit activities or investigations or findings. • Working documents involving confidential information must be properly safeguarded.
5.3 Executing the Audit	
5.3.1 Opening Meeting	Arrange an opening meeting to: • Introduce the audit team • Review the audit plan, scope, and objectives • Provide a short summary of audit procedures to the auditee • Establish formal communication between the audit team and the auditee • Confirm the availability of adequate resources and facilities to conduct the audit • Confirm the audit schedule • Clarify and confirm the details of the audit plan
5.3.2 Examination	
5.3.2.1 Collecting Evidence	• Evidence should be collected through interviews, examination of documents, and observation of activities and conditions and should be validated through similar information gathered from other independent sources, such as physical observation, measurements, and records. • Changes in the audit plan and procedures should be made, as necessary, to ensure optimal achievement of audit objectives.
5.3.2.2 Audit Observations	• Audit observations should be documented and reviewed to identify nonconformances.

System Elements	Description
	• Findings should be supported by objective evidence.
	• Audit findings should be reviewed by the lead auditor with the responsible manager.
5.3.3 Closing meeting with Auditee	• At the end of the audit, and before preparing the audit report, the audit findings should be discussed with the auditee management to ensure that they clearly understand the audit results.
	• Records of closing meeting are to be kept.
5.4 Audit Documents	
5.4.1 Audit Report Preparations	• Audit report is prepared under the direction of the lead auditor, who is responsible for its accuracy and completeness.
5.4.2 Report Content	• The audit report should be formal in style, format, and content. It should be dated and signed by the lead auditor. • The report should include: – scope and objective of the audit – details of audit plan – audit schedule – audit team and auditee representatives – reference documents and standards against which the audit was conducted – observation of nonconformances – compliance status of the quality system – system performance in achieving defined quality objectives – audit report distribution list
5.4.3 Report Distribution	• Audit report to be sent to client by the lead auditor. • Client to provide a copy of the report to the auditee's management.

System Elements	Description
	• Further distribution of audit report should be done in consultation with the auditor. • Confidential information should be safeguarded. • Audit report should be issued as soon as possible after the completion of the audit.
5.4.4 Record Retention	• Auditing documents should be retained by agreement between the client, auditing organization, and the auditee, and in accordance with any regulatory requirements.
6. **Audit Completion**	• Audit is deemed to be completed upon submission of the audit report to the client.
7. **Corrective Action Follow-up**	• Auditee is responsible for taking corrective action to correct identified nonconformities. • Corrective action and subsequent follow-up audits should be completed within a time period agreed to by the client and the auditee in consultation with the auditing organization.

THE HUMAN FACTOR IN AUDITING

The preceding sections provide sufficient guidelines on the principles and procedure for conducting an environmental quality audit. However, the success of an audit is totally dependent on the ability of the auditors. It is important, therefore, to elaborate further on the role and contribution of human aspects in the process of auditing.

To begin with, an internal audit and an external audit take place in very different environments. The objective of internal auditing is to reduce costs and improve quality and customer satisfaction through self–evaluation of the quality system. The atmosphere is a bit more relaxed because all the personnel and functions involved in the exercise are part of the same organization and all of them share a common goal of system improvement. External audits, on the other hand, are performed by an auditing organization external to the company, either on behalf of the customer or for the purpose of certification to the quality system requirements of a standard. The atmosphere in external auditing is generally one of apprehension and anxiety.

Notwithstanding the type of audit or its environment, one thing is evident—it is the ability of the auditor that bears the greatest influence on the effectiveness and outcome of the audit. For any auditing program to be successful, it is imperative to give special considerations to the behavioral and human aspects of auditing. In this section, we shall elaborate on these three aspects:

1. Personal characteristics of effective auditors

2. Procedural do's and don'ts for the auditors

3. Different personality types auditors may encounter

Personal Characteristics of an Effective Auditor

Personal Attributes
- Decent outward physical appearance

- Intelligent, constructive, proactive

- Helpful, patient, mature, honest, reliable, stable

- Careful, diplomatic, objective, unbiased, self–confident

- Task–oriented, persistent, tenacious, consistent, observant

- Participative, cooperative, team–oriented

Work Skills

- Adequate education, practical experience, and subject–matter knowledge

- Proper training in auditing procedures

- Appropriate product and process knowledge

- Analytical ability to make objective decisions

- Ability to be decisive, precise, and judicious

- Ability to be independent, systematic, and autonomous

- Capability to adapt to changing conditions and assignments

Management Skills

- Planning and organizational skills

- Leadership abilities

- Ability to supervise and delegate

- Ability to control and direct operations

- Reporting, record–keeping, and budgeting skills

- Ability to manage teamwork

- Avoidance of conflict of interest

Communication Skills

- Ability to listen, understand, and respond effectively

- Clarity of expression: verbal and written

- Ability to communicate effectively

- Ability to gain cooperation of all parties: client, auditee, and audit team

- Honesty and integrity in dealing with all personnel

- Conformance to agreed–upon procedures and rules
- Respect for confidentiality
- Objectivity in reporting results

Listening Skills
- Must be an adept listener
- Ability to observe body language
- Ability to listen to the speaker without interruption
- Ability to maintain concentration
- Must not prejudge the speaker
- Must not react emotionally
- Ability to differentiate between important and trivial details
- Ability to effectively highlight the main points of discussion

Procedural Do's and Don'ts for the Auditors

- During the conduct of an audit, the auditor must be:
    - helpful, approachable, tolerant, respectful, considerate, cooperative, considerate
    - attentive, assertive, rational, specific, resolute, consistent, factual, explicit, thorough, persistent
    - diplomatic, unbiased, discreet, objective, honest
    - patient, calm, diligent, polite, adaptable, brief
- The auditor should avoid appearing:
    - arrogant, overbearing, unfriendly, stubborn, egotistic
    - careless, vague, sloppy, nitpicky, uncertain, ambiguous

– emotional, tactless, narrow-minded, prejudiced, deceptive, impatient, rude, sneaky, antagonistic

– unfair, thoughtless, unreasonable, conflicting

Personality Types Auditors May Encounter

During the conduct of an audit, the auditor is going to meet many different personality types at the auditee organization. Different people react in different ways during the audit. It is important for the auditor to understand this phenomenon and seek objectivity out of this indiscriminate variety. If care is not exercised, the auditor may end up either wasting time or being swayed to make faulty conclusions. Following are some of the types of persons or responses that the auditor can encounter during the audit.

Arrogant or Hostile: Some persons will display outright hostility whenever the auditor identifies any system problem or nonconformance. They do not like the auditor to tell them how to run their operation. In their judgment, they have been successfully producing quality products and services all along and consequently there is no possibility of error in their function. Normally, these types of persons fail to understand the difference between product inspection and quality system audit. With these types of persons, the auditor has to be very patient while at the same time being specific, rational, factual, and assertive.

Resistant: The auditor may encounter a person who is convinced that the system is excellent and there are no problems. The auditor must be persistent and diplomatic with such persons while being diligently objective.

Ignorant or Ill–Prepared: Some people may, intentionally or unintentionally, display lack of operational knowledge in answering the question and pass the buck onto either other functions or the management whenever the auditor needs a clarification. The auditor has to be very careful in such cases, because it can either waste time or result in insufficient evidence to justify audit findings. The auditor must be diligent and diplomatic, but astutely assertive and demanding. The auditor may ultimately have to ask the auditee management for a replacement if the situation persists.

Overly Detailed: In contrast to the above situation, the auditor may encounter a person who possesses exquisite knowledge of all the functions under his command. Such a person would like to clarify every minute detail to the auditor in a grandiose manner which may result in a total disruption of the audit schedule and plan. While the auditor should express appreciation of the person's knowledge, it is important to be explicit, specific, and unyielding in showing objectivity and discretion.

Carefree or Indifferent: The auditor may encounter the carefree personality–type for whom nothing matters. Such auditees think that though they shall take into consideration the auditor's findings, it is not going to make any difference in the product quality. The auditor must be consistently thorough and task–minded with such a person.

Argumentative: The auditor may encounter a fighter, who would fight back and argue on everything whether it makes sense or not. In such situations, the auditor should try to be relevant, factual, and diplomatically assertive.

Overly Judicial: Some persons may be overly formal and judicial in their attitude and want to follow the audit strictly as per the stipulated contractual rules and procedures. The situation can easily turn into a courtroom drama creating an uneasy environment for the auditor. While respecting the protocol to the extent possible or feasible, the auditor has to handle the situation tactfully and patiently. However, the auditor must be persistent in performing the job dutifully, thoroughly, and objectively.

A Final Word

It is a long road from an organization's mission statement and commitment to becoming environmentally conscientious to implementing an effective EMS and receiving ISO 14000 certification through a third-party registrar. But the process does not end with certification. To maintain certification in both ISO 14000 and ISO 9000, an organization must continuously review, maintain, and upgrade. TQM also requires continuous maintenance to be effective. Reviewing, maintaining, and upgrading all these systems entails a tremendous effort on the part of an organization, but it is an effort that will continuously reap great rewards—providing top quality products, improving the quality of the environment, and enhancing competitiveness. These are, after all, the core elements of the new paradigm of success—an integrated approach giving equal attention to quality management of products/services and quality management of environmental impacts. This new paradigm is a strategy that will help position an organization for competitive success in the year 2000 and beyond. One viable way to get on and stay on this road to success is for an organization to be certified in ISO 9000 as well as ISO 14000. We hope this book has assisted you in finding that road, as well as in fulfilling the very important mission of contributing to a better environment.

In Chapter 4, we discussed in some detail the British Standard, BS 7750, and the Canadian Standard, CSA-Z750, on the subject of environmental protection. We also gave an outline of the South African Standard, SABS-02510. Here, we will present twelve important international protocols, listed below.

- The Rio Declaration on Environment and Development: The United Nations

- International Chamber of Commerce (ICC): Business Charter for Sustainable Development

- European Petroleum Industry Association (EUROPIA): Environmental Guiding Principles

- Keidanren (Japan Federation of Economic Organizations): Keidanren Global Environment Charter

- Coalition for Environmentally Responsible Economies (CERES) Principles (formerly the Valdez Principles)

- Business Council on National Issues: Business Principles for a Sustainable and Competitive Future

- National Round Table on the Environment and the Economy (NRTEE): Objectives for Sustainable Development

- Canadian Chemical Producers' Association (CCPA): Responsible Care—Guiding Principles

- The Environmental Policy of the Coca-Cola Company

- The Environmental Policy of British Telecom

- The Environmental Policy of National Power

- World Industry Council for the Environment (WICE): Guidelines on Environmental Reporting

Since we are only presenting a general framework of these protocols, the reader is advised to consult the source documents for greater details.

There is an enormous amount of useful information contained in these protocols that an organization can gainfully utilize in many different ways such as:

- Developing environmental mission statement, policy, and environmental objectives

- Developing a set of core environmental values and principles

- Generating environmental quality management system elements

- Establishing and implementing the requisite environmental quality specifications and requirements

THE RIO DECLARATION ON ENVIRONMENT AND DEVELOPMENT

The United Nations Conference on Environment and Development, having met at Rio de Janeiro from 3 to 14 June 1992, reaffirming the Declaration of the United Nations Conference on the Human Environment, adopted at Stockholm on 16 June 1972, and seeking to build upon it, with the goal of establishing a new and equitable global partnership through the creation of new levels of cooperation among States, key sectors of societies and people, working towards international agreements which respect the interests of all and protect the integrity of the global environmental and developmental system, recognizing the integral and interdependent nature of the Earth, our home proclaims that:

Principle 1

Human beings are at the center of concerns for sustainable development. They are entitled to a healthy and productive life in harmony with nature.

Principle 2

States have, in accordance with the Charter of the United Nations and the principles of international law, the sovereign right to exploit their own resources pursuant to their own environmental and developmental policies, and the responsibility to ensure that activities within their jurisdiction or control do not cause damage to the environment of other States or of areas beyond the limits of national jurisdiction.

Principle 3

The right to development must be fulfilled so as to equitably meet the developmental and environmental needs of present and future generations.

Principle 4

In order to achieve sustainable development, environmental protection shall constitute an integral part of the development process and cannot be considered in isolation from it.

Principle 5

All States and all people shall cooperate in the essential task of eradicating poverty as an indispensable requirement for sustainable development, in order to decrease the disparities in standards of living and better meet the needs of the majority of the people of the world.

Principle 6

The special situation and needs of developing countries, particularly the least developed and those most environmentally vulnerable, shall be given special priority. International actions in the field of environment and development should also address the interest and needs of all countries.

Principle 7

States shall cooperate in a spirit of global partnership to conserve, protect, and restore the health and integrity of the Earth's ecosystem. In view of the different contributions to global environmental degradation, States have common but differentiated responsibilities. The developed countries acknowledge the responsibility that they bear in the international pursuit of sustainable development in view of the pressures their societies place on the global environment and of the technologies and financial resources they command.

Principle 8

To achieve sustainable development and a higher quality of life for all people, States should reduce and eliminate unsustainable patterns of production and consumption, and promote appropriate demographic policies.

Principle 9

States should cooperate to strengthen endogenous capacity-building for sustainable development by improving scientific understanding through exchanges of scientific and technological knowledge, and by enhancing the development, adaptation, diffusion, and transfer of technologies, including new and innovative technologies.

Principle 10

Environmental issues are best handled with the participation of all concerned citizens, at the relevant level. At the national level, each individual shall have appropriate access to information concerning the environment that is held by public authorities, including information on hazardous materials and activities in their communities, and the opportunity to participate in decision-making processes. States shall facilitate and encourage public awareness and participation by making information widely available. Effective access to judicial and administrative proceedings, including redress and remedy, shall be provided.

Principle 11

States shall enact effective environmental legislation. Environmental standards, management objectives, and priorities should reflect the environmental and developmental context to which they apply. Standards applied by some countries may be inappropriate and of unwarranted economic and social cost to other countries, in particular developing countries.

Principle 12

States should cooperate to promote a supportive and open international economic system that would lead to economic growth and sustainable development in all countries, to better address the problems of environmental degradation. Trade policy measures for environmental purposes should not constitute a means of arbitrary or unjustifiable discrimination or a disguised restriction on international trade. Unilateral actions to deal with environmental challenges outside the jurisdiction of the importing country should be avoided. Environmental measures addressing transboundary or global environmental problems should, as far as possible, be based on an international consensus.

Principle 13

States shall develop national law regarding liability and compensation for the victims of pollution and other environmental damage. States shall also cooperate in an expeditious and more determined manner to develop further international law regarding liability and compensation for adverse effects of environmental damage caused by activities within their jurisdiction or control to areas beyond their jurisdiction.

Principle 14

States should effectively cooperate to discourage or prevent the relocation and transfer to other States of any activities and substances that cause severe environmental degradation or are found to be harmful to human health.

Principle 15

In order to protect the environment, the precautionary approach shall be widely applied by States according to their capabilities. Where there are threats of serious or irreversible damage, lack of full scientific certainty shall not be used as a reason for postponing cost-effective measures to prevent environmental degradation.

Principle 16

National authorities should endeavor to promote the internalization of environmental costs and the use of economic instruments, taking into account the approach that the polluter should, in principle, bear the cost of pollution, with due regard to the public interest and without distorting international trade and investment.

Principle 17

Environmental impact assessment, as a national instrument, shall be undertaken for proposed activities that are likely to have a significant adverse impact on the environment and are subject to a decision of a competent national authority.

Principle 18

States shall immediately notify other States of any natural disasters or other emergencies that are likely to produce sudden harmful effects on the environment of those States. Every effort shall be made by the international community to help States so afflicted.

Principle 19

States shall provide prior and timely notification and relevant information to potentially affected States on activities that may have a significant adverse transboundary environmental effect and shall consult with States at an early stage and in good faith.

Principle 20

Women have a vital role in environmental management and development. Their full participation is therefore essential to achieve sustainable development.

Principle 21

The creativity, ideals, and courage of the youth of the world should be mobilized to forge a global partnership in order to achieve sustainable development and ensure a better future for all.

Principle 22

Indigenous people and their communities, and other local communities, have a vital role in environmental management and development because of their knowledge and traditional practices. States should recognize and duly support their identity, culture, and interest, and enable their effective participation in the achievement of sustainable development.

Principle 23

The environment and natural resources of people under oppression, domination, and occupation shall be protected.

Principle 24

Warfare is inherently destructive of sustainable development. States shall therefore respect international law providing protection for the environment in times of armed conflict and cooperate in its further development, as necessary.

Principle 25

Peace, development, and environmental protection are interdependent and indivisible.

Principle 26

States shall resolve all their environmental disputes peacefully and by appropriate means in accordance with the Charter of the United Nations.

Principle 27

States and people shall cooperate in good faith and in a spirit of partnership in the fulfillment of the principles embodied in this Declaration and in the further development of international law in the field of sustainable development.

INTERNATIONAL CHAMBER OF COMMERCE

Business Charter for Sustainable Development

1. Corporate Policy. To recognize environmental management as among the highest corporate priorities and as a key determinant to sustainable development; to establish policies, programs, and practices for conducting operations in an environmentally sound manner.

2. Integrated Management. To integrate these policies, programs, and practices fully into each business as an essential element of management in all its functions.

3. Process of Improvement. To continue to improve policies, programs, and environmental performance, taking into account technical developments, scientific understanding, consumer needs, and community expectations, with legal regulations as starting point; and to apply the same environmental criteria internationally.

4. Employee Education. To educate, train, and motivate employees to conduct their activities in an environmentally responsible manner.

5. Prior Assessment. To assess environmental impacts before starting a new activity or project and before decommissioning a facility or leaving a site.

6. Products and Services. To develop and provide products or services that have no undue environmental impact and are safe in their intended use, that are efficient in their consumption of energy and natural resources, and that can be recycled, reused, or disposed of safely.

7. Customer Advice. To advise, and where relevant educate, customers, distributors, and the public in the safe use, transportation, storage, and disposal of products provided; and to apply similar considerations to the provisions of services.

8. Facilities and Operations. To develop, design, and operate facilities and conduct activities taking into consideration the efficient use of energy and materials, the sustainable use of renewable resources, the minimization of adverse environmental impact and waste generation, and the safe and responsible disposal of residual wastes.

9. Research. To conduct or support research on the environmental impacts of raw materials, products, processes, emissions, and wastes associated with the enterprise and in the means of minimizing such adverse impacts.

10. Precautionary Approach. To modify the manufacture, marketing, or use of products or services or the conduct of activities, consistent with scientific and technical understanding, to prevent serious or irreversible environmental degradation.

11. Contractors and Suppliers. To promote the adoption of these principles by contractors acting on behalf of the enterprise, encouraging and, where appropriate, requiring improvements in their practices to make them consistent with those of the enterprise; and to encourage the wider adoption of these principles by suppliers.

12. Emergency Preparedness. To develop and maintain, where significant hazards exist, emergency preparedness plans in conjunction with the emergency services, relevant authorities, and the local community, recognizing potential transboundary impacts.

13. Transfer of Technology. To contribute to the transfer of environmentally sound technology and management methods throughout the industrial and public sectors.

14. Contributing to the Common Effect. To contribute to the development of public policy and to business, governmental, and intergovernmental programs and educational initiatives that will enhance environmental awareness and protection.

15. Openness to Concerns. To foster openness and dialogue with employees and the public, anticipating and responding to their concerns about potential hazards and impacts of operations, products, wastes, or services, including those of transboundary or global significance.

16. Compliance and Reporting. To measure environmental performance; to conduct regular environmental audits and assessments of compliance with other company requirements, legal requirements, and these principles; and periodically to provide appropriate information to the board of Directors, shareholders, employees, the authorities, and the public.

EUROPEAN PETROLEUM INDUSTRY ASSOCIATION (EUROPIA)

Environmental Guiding Principles

1. Make the principles set forth herein a high priority in the definition and implementation of corporate strategies.

2. Adapt where necessary internal procedures, industry practices, and other operating guidelines towards the goal of protecting the environment and the health and safety of individuals.

3. Conduct operations and handle raw materials and products in a manner that protects the environment and the health and safety of employees and the public, while conserving natural resources and using energy efficiently.

4. Develop and maintain procedures to reduce the risk of spills or accidental emissions; maintain appropriate emergency response procedures in case of accidents.

5. Develop programs to reduce overall emissions and waste generation.

6. Ensure that adequate waste management programs are developed and carried out, which will allow the disposal of wastes as safely as is reasonably practicable.

7. Work with others to resolve problems arising out of the handling and disposal of hazardous substances from members' operations.

8. Provide advice to customers, contractors, or others on the safe use, handling, transportation, and disposal of raw materials, products, and wastes from members' operations.

9. Inform appropriate officials, employees, customers, and the public in a timely manner on significant industry-related safety, health, and environmental hazards, and recommend protective measures.

10. Support research and development programs to study the effects of the industry's activities on the environment, the health and safety of individuals, and the prevention of the risks connected hereto.

11. Promote among employees an individual and collective sense of responsibility for the preservation of the environment and protection of health and safety of individuals.

12. Work and consult with authorities drafting laws, regulations or procedures to safeguard the community, workplace and environment.

13. Promote these principles and practices by sharing experiences and offering technical assistance to others who deal with similar raw materials, petroleum products, and wastes.

KEIDANREN (JAPAN FEDERATION OF ECONOMIC ORGANIZATIONS)

Keidanren Global Environment Charter

Basic Philosophy

A company's existence is closely bound up with the global environment as well as with the community it is based in. In carrying on its activities each company must maintain respect for human dignity, and strive toward a future society where the global environment is protected. We must aim to construct a society whose members cooperate together on environmental problems, a society where sustainable development on a global scale is possible, where companies enjoy a relationship of mutual trust with local citizens and consumers, and where they vigorously and freely develop their operations while preserving the environment. Each company must aim at being a good global corporate citizen, recognizing that grappling with environmental problems is essential to its own existence and its activities.

Guidelines for Corporate Action

1. General Management Principles

Companies should always carry on their business activities to contribute to the establishment of a new economic social system for realizing an environmentally protective society leading to sustainable development.

2. Corporate Organization

(a) Companies shall establish an internal system to handle environmental issues by appointing an executive and creating an organization in charge of environmental problems.

(b) Environmental regulation shall be established for company activities, and these shall be observed. Such internal regulations shall include goals for reducing the load on the environment. An internal inspection to determine how well the environmental regulations are being adhered to shall be carried out at least once a year.

3. Concern for the Environment

(a) All company activities, beginning with siting of production facilities, shall be scientifically evaluated for their impact on the environment, and necessary counter-measures shall be implemented.

(b) Care shall be taken in the research, design, and development stages of making a product to lessen the possible burden on the environment at each level of its production, distribution, appropriate use, and disposal.

(c) Companies shall strictly observe all national and local laws and regulations of environmental protection, and where necessary they shall set additional standards of their own.

(d) When procuring materials, including materials for production, companies shall endeavor to purchase those that are superior for conserving resources, preserving the environment, and enhancing recycling.

(e) Companies shall employ technologies that allow efficient use of energy and preservation of the environment in their production and other activities. Companies shall endeavor to use resources efficiently and reduce waste products through recycling, and shall appropriately deal with pollutants and waste products.

4. Technology Development

In order to help solve global environmental problems, companies shall endeavor to develop and supply innovative technologies, products, and services that allow conservation of energy and other resources together with preservation of the environment.

5. Technology Transfers

(a) Companies shall seek appropriate means for the domestic and overseas transfer of their technologies, know-how, and expertise for dealing with environmental problems and conserving energy and other resources.

(b) In participating in official development assistance projects, companies shall carefully consider environmental and anti-pollution measures.

6. Emergency Measures

(a) If environmental problems ever occur as a result of an accident in the course of company activities or deficiency in a product, companies shall adequately explain the situation to all concerned parties and take appropriate measures, using their technologies and human and other resources to minimize the impact on the environment.

(b) Even when a major disaster or environmental accident occurs outside of a company's responsibility, it shall still actively provide technological and other appropriate assistance.

7. Public Relations and Education

(a) Companies shall actively publicize information and carry out educational activities concerning their measures for protecting the environment, maintaining ecosystems, and ensuring health and safety in their activities.

(b) The employees shall be educated to understand the importance of daily close management to ensure the prevention of pollution and the conservation of energy and other resources.

(c) Companies shall provide users with information about the appropriate use and disposal, including recycling, of their products.

8. Community Relations

(a) As community members, companies shall actively participate in activities to preserve the community environment and support employees who engage in such activities on their own initiative.

(b) Companies shall promote dialogue with people in all segments of society over operational issues and problems seeking to achieve mutual understanding and strengthen cooperative relations.

9. Overseas Operations

Companies developing operations overseas shall observe the Ten-Points—Environmental Guidelines for the Japanese Enterprises Operating Abroad in Keidanren's Basic Views of the Global Environmental Problems.

10. Contribution to Public Policies

(a) Companies shall work to provide information gained from their experiences to administrative authorities, international organizations, and other bodies formulating environmental policy, as well as participate in dialogue with such bodies, in order that more rational and effective policies can be formulated.

(b) Companies shall draw on their experience to propose rational systems to administrative authorities and international organizations concerning formulation of environmental policies and to offer sensible advice to consumers on lifestyles.

11. Response to Global Problems
(a) Companies shall cooperate in scientific research on the causes and effects of such problems as global warming and they shall also cooperate in the economic analysis of possible counter-measures.
(b) Companies shall actively work to implement effective and rational measures to conserve energy and other resources even when such environmental problems have not been fully elucidated by science.
(c) Companies shall play an active role when the private sector's help is sought to implement international environmental measures, including work to solve the problems of poverty and overpopulation in developing countries.

COALITION FOR ENVIRONMENTALLY RESPONSIBLE ECONOMIES (CERES) PRINCIPLES (FORMERLY THE VALDEZ PRINCIPLES)

Protection of the Biosphere. We will reduce and make continual progress toward eliminating the release of any substance that may cause environmental damage to the air, water, or the earth or its inhabitants. We will safeguard all habitats affected by our operations and will protect open spaces and wilderness, while preserving biodiversity.

Sustainable Use of Natural Resources. We will make sustainable use of renewable natural resources, such as water, soils, and forests. We will conserve nonrenewable natural resources through efficient use and careful planning.

Reduction and Disposal of Waste. We will reduce and where possible eliminate waste through source reduction and recycling.

All waste will be handled and disposed of through safe and responsible methods.

Energy Conservation. We will conserve energy and improve the energy efficiency of our internal operations and of the goods and services we sell. We will make every effort to use environmentally safe and sustainable energy sources.

Risk Reduction. We will strive to minimize the environmental, health, and safety risks to our employees and the communities in which we operate through safe technologies, facilities, and operating procedures, and by being prepared for emergencies.

Safe Products and Services. We will reduce and where possible eliminate the use, manufacture, or sale of products and services that cause environmental damage or health or safety hazards. We will inform our customers of the environmental impacts of our products or services and try to correct unsafe use.

Environmental Restoration. We will promptly and responsibly correct conditions we have caused that endanger health, safety, or the environment. To the extent feasible, we will redress injuries we have caused to persons or damage we have caused to the environment and will restore the environment.

Informing the Public. We will inform in a timely manner everyone who may be affected by conditions caused by our company that might endanger health, safety, or the environment. We will regularly seek advice and counsel through dialogue with persons in communities near our facilities. We will not take any action against employees for reporting dangerous incidents or conditions to management or to appropriate authorities.

Management Commitment. We will implement these principles and sustain a process that ensures that the Board of Directors and Chief Executive Officer are fully informed about pertinent environmental issues and are fully responsible for environmental policy. In selecting our Board of Directors, we will consider demonstrated environmental commitment as a factor.

Audits and Reports. We will conduct an annual self-evaluation of our progress in implementing these principles. We will support the timely creation of generally accepted environmental audit procedures. We will annually complete the CERES Report, which will be made available to the public.

Disclaimer. These principles establish an environmental ethic with criteria by which investors and others can assess the environmental performance of companies. Companies that sign these principles pledge to go voluntarily beyond the requirements of the law. These principles are not intended to create new legal liabilities, expand existing rights or obligations, waive legal defenses, or otherwise affect the legal position of any signatory company, and are not intended to be used against a signatory in any legal proceeding for any purpose.

BUSINESS COUNCIL ON NATIONAL ISSUES

Business Principles for a Sustainable and Competitive Future

1. Adopt sustainable development as a key operating principle of the company. Under senior management direction, and utilizing appropriate employee training and motivation, develop corporate systems and procedures to ensure the company operates according to the principles of sustainable development.

2. Develop corporate goals and objectives for sustainable development, and a means to measure progress against these objectives. Communicate periodically to the board, shareholders, employees, government authorities, and the public with respect to these goals and progress made.

3. Promote public policies and regulatory frameworks within which market forces can be fully responsive to the choices of individuals and organizations in working towards sustainable development.

4. Meet or exceed all applicable environmental laws, regulations, and standards. Set the company's own standards when government regulations do not exist.

5. Before launching any new project, product, or service, undertake an evaluation of its sustainability, and integrate into the planning process measures to prevent or minimize any potential environmental impact.

6. Adopt the principle of life cycle management by applying sustainability criteria at every stage of the enterprise's activity—from R&D, design for recycling and reuse, and the utilization of raw materials and hazardous substances, to production processes, transportation and distribution, sales and customer use, and ultimate disposal.

7. Take a proactive role in promoting the goal of sustainable development, both nationally and internationally, and work cooperatively with government, labor, and public interest groups to develop policies to promote sustainable development practices by suppliers, customers, and others in the business community.

8. Consider means to facilitate the transfer of environmentally beneficial technologies, throughout the business sector and internationally, by the deployment of managerial, technical, and financial resources.

NATIONAL ROUND TABLE ON THE ENVIRONMENT AND THE ECONOMY (NRTEE)

Objectives for Sustainable Development

1. Stewardship. We must preserve the capacity of the biosphere to evolve by managing our social and economic activities for the benefit of present and future generations.

2. Shared Responsibility. Everyone shares the responsibility for a sustainable society. All sectors must work towards this common purpose, with each being accountable for its decisions and actions, in a spirit of partnership and open cooperation.

3. Prevention and Resilience. We must try to anticipate and prevent future problems by avoiding the negative environmental, economic, social, and cultural impacts of policy, programs, decisions, and development activities. Recognizing that there will always be environmental and other events which we cannot anticipate, we should also strive to increase social, economic, and environmental resilience in the face of change.

4. Conservation. We must maintain and enhance essential ecological processes, biological diversity, and life support systems of our environment and natural resources.

5. Energy and Resource Management. Overall, we must reduce the energy and resource content of growth, harvest renewable resources on a sustainable basis, and make wise and efficient use of our nonrenewable resources.

6. Waste Management. We must first endeavor to reduce the production of waste, then reuse, recycle, and recover waste by-products of our industrial and domestic activities.

7. Rehabilitation and Reclamation. Our future policies, programs, and development must endeavor to rehabilitate and reclaim damaged environments.

8. Scientific and Technological Innovation. We must support education and research and development of technologies, goods, and services essential to maintaining environmental quality, social and cultural values, and economic growth.

9. International Responsibility. We must think globally when we act locally. Global responsibility requires ecological interdependence among provinces and nations, and an obligation to accelerate the integration of environmental, social, cultural, and economic goals. By working cooperatively within Canada and internationally, we can develop comprehensive and equitable solutions to problems.

10. Global Development. Canada should support methods that are consistent with the preceding objectives when assisting developing nations.

CANADIAN CHEMICAL PRODUCERS' ASSOCIATION (CCPA)

Responsible Care: Guiding Principles

Each member company has subscribed to the following guiding principles:

- Ensure that its operations do not present an unacceptable level of risk to employees, customers, the public, or the environment.

- Provide relevant information on the hazards of chemicals to its customers, urging them to use and dispose of products in a safe manner, and make such information available to the public on request.

- Make responsible care an early and integral part of the planning process leading to new products, processes, or plants.

- Increase the emphasis on the understanding of existing products and their uses, and ensure that a high level of understanding of new products and their potential hazards is achieved prior to and throughout commercial development.

- Comply with all legal requirements which affect its operations and products.

- Be responsive and sensitive to legitimate community concerns.

- Work actively with and assist governments and selected organizations to foster and encourage equitable and attainable standards.

- These principles are supplemented with more specific principles for each specific area of concern.

THE ENVIRONMENTAL POLICY OF THE COCA-COLA COMPANY

The Coca-Cola Company is built on a few remarkably simple ideas: refreshment, quality, service. Because of our commitment to these principles, we have become the world's preeminent soft drink company, with consumers in nearly 200 countries around the world.

The Coca-Cola Company is also built on the idea of responsible corporate citizenship. Directly and indirectly, we touch the lives of billions of people around the world, and our responsibility to them includes conducting our business in ways that protect and preserve the environment.

In fulfilling this responsibility, the Coca-Cola Company and its subsidiaries adhere to the following fundamental principles:

We conduct our operations in compliance with applicable environmental laws and regulations. Even in the absence of governmental regulation, we operate in an environmentally responsible manner.

We minimize the environmental impact of our operations, products, and packages through research and the application of new technology.

We minimize the discharge of waste materials into the environment by utilizing responsible pollution control practices.

We recognize the interrelationship between energy and the environment, and we promote the efficient use of energy throughout our system.

We support efforts to understand and address the problems of solid waste management. We are committed to both reducing and recycling the solid waste generated in our own facilities and to helping communities where we operate implement recycling and sound solid waste management systems.

We believe in the concept of environmental accountability and conduct periodic audits of our performance and practices. We share nonproprietary information about our progress with the public.

We encourage and participate responsibly in discussion of environmental concerns. We cooperate with public and government organizations in seeking solutions to environmental problems.

This environmental commitment permeates our organization and our officers, managers, and employees assume responsibility for its daily implementation.

THE ENVIRONMENTAL POLICY OF BRITISH TELECOM

BT is committed to minimizing the impact of its operations on the environment by means of a program of continuous improvement.

In particular BT will:

- Meet and, where appropriate, exceed the requirements of all relevant legislation—where no regulations exist BT shall set its own exacting standards.

- Seek to reduce consumption of materials in all operations, reuse rather than dispose whenever possible, and promote recycling and the use of recycled materials.

- Design energy efficiency into new services, buildings, and products, and manage energy wisely in all operations.

- Reduce wherever practicable the level of harmful emissions.

- Market products that are safe to use, and make efficient use of resources, especially those which can be reused, recycled, or disposed of safely.

- Work with its suppliers to minimize the impact of their operations on the environment through a quality purchasing policy.

- Site its buildings, structures, and operational plant to minimize visual, noise, and other impacts on the local environment.

- Support through its community program the promotion of environmental protection by relevant external groups and organizations.

- Include environmental issues in discussions with the BT unions and the BT training programs, and encourage the implementation by all BT people of sound environmental practices.

- Monitor progress and publish an environmental performance report on an annual basis.

THE ENVIRONMENTAL POLICY OF NATIONAL POWER

Our environmental policy was revised and strengthened in 1992. Five policy principles guide our activities. These ensure that environmental considerations are an important part of whatever we do, in the UK and overseas:

Approach. To integrate environmental factors into business decisions.

Compliance. To monitor compliance with environmental regulations and to perform better than they require, where appropriate.

Improvement. To improve our environmental performance continuously.

Accountability. To review regularly at Board level, and to make public, the company's environmental performance.

Responsibility. To establish a reputation for effective environmental management.

WORLD INDUSTRY COUNCIL FOR THE ENVIRONMENT (WICE)

Guidelines on Environmental Reporting

The World Industry Council for the Environment (WICE) is a global coalition of enterprises initiated in 1993 by the International Chamber of Commerce (ICC). Its membership includes enterprises representing a wide diversity of sectors from countries inside and outside the OECD.

1. Foreword by a Senior Responsible Person. This provides an opportunity to demonstrate commitment from the top. This is important to both external audiences and employees.

2. Profile of Enterprise. The size, activity, and location of operations, described in a way that will help the reader put the rest of the report in context.

3. Environmental Policy. Your environmental policy provides an important basis for reporting. The ICC Business Charter or the guidelines your enterprise may support can be used to develop or update the policy.

4. Environmental Targets and Objectives. Published targets or objectives are frequently the driving force behind continuous improvement in environmental performance. When establishing these objectives, be aware of the potential costs and possible legal implications.

5. Views on Environmental Issues. Provide information on public issues that are or could be relevant to the activities of your enterprise, in order to contribute to the public debate.

6. Community Relations. Provide evidence of your concern for your staff and of good citizenship.

7. Environmental Management Systems. An environmental report may emphasize management programs and practices used to implement the policy; the organization of environmental responsibility and expertise; and the education, training and motivation of employees. An environmental report can provide information on specific management programs. This provides an opportunity to demonstrate how your environmental policy is translated into action. (The ICC Business Charter, principles 1-4, 11, 16.)

8. Management of Environmental Risks. As part of environmental management, describe the action being taken to identify and minimize environmental risks. You may wish to include information on past incidents and the action taken to prevent a recurrence.

9. Office and Site Practices. The positive impact of environmental management is proven in good office practices, such as recycling and energy management.

10. Environmental Indicators and Targets. A report may contain data on emissions, effluents and discharges to air, water, and soil, and provide information on specific local community concerns such as noise and smells. Data should be presented in context; for example, business performance over previous years. Degree of compliance to regulatory requirements or comparable data on different sites should be included.

11. Use of Energy and Natural Resources. These include conservation measures being taken across all activities: in the office, factory and transportation.

12. Compliance with Regulations. Enterprises may include details on the level of compliance. This can be based on information provided to the authorities. An effective management system encourages compliance as a minimum. In areas where regulations are less rigorous, enterprises might seek to explain how their performance meets their own environmental standards.

13. Financial Indicators. Some enterprises, either because of the type of activity or set of products and services, use financial data as an indicator of their environmental performance. This may include liabilities, provisions for future remediation, and US Securities and Exchange Commission (or equivalent) reportable data. Fund managers are beginning to look regularly to this data as part of their decision making.

14. Products, Processes, and Services. This may include information on what the enterprise is doing to improve the environmental performance of its products and services. Examples may include:

- product design and life cycle assessment (LCA)

- pollution prevention

- auditing

15. Giving more information. The report may provide contact names to allow readers to make comments or obtain further information about your environmental programs.

In Chapter 1, we briefly touched upon the basic premise of establishing a quality management system to improve the quality of products, services, and work-life. Quality improvement does not come in a ready-made package, nor is it a separate planning activity. That is why we urge companies who are at the threshold of revolutionizing their organizations through implementing a comprehensive environmental management system to consider teaming that effort with the equally wide-ranging strategic implementation of a total quality management (TQM) program. Quality is not a tactical but a strategic issue, whether it is quality of products and services, or quality of environmental impact. Everything—quality philosophies, practices, tools, techniques, and training—must be embedded in the business plan. There are layers of activities that need to be undertaken to address quality improvement concerns.

THE QUALITY SYSTEM

Quality does not come through piecemeal efforts or through a single quality improvement program, procedure, or process. Quality is the result of a totally integrated set of actions with a long-term commitment. The balance of this appendix gives detailed guidelines on how to develop, implement, and maintain a quality system. (For a more in-depth treatment of TQM, the reader can refer to the author's book, *ISO 9000 Certification and Total Quality Management.*)

A TQM model can be implemented in almost any conceivable situation. However, the magnitude and extent of the TQM exercise would depend upon the nature of the enterprise, its size and complexity, the type of activity, quality requirements, financial and resource base, clientele, etc. Consequently, the TQM model and its implementation has to be commensurate with the overall profile of the organization. Basically there are three distinct organization types with their own unique requirements:

Manufacturing Organizations
- Tangible product production/assembly

- Servicing

- Customer satisfaction

- Marketability/profitability

Service Organizations
- Provision of services, with or without tangible products

- Customer satisfaction

- Profitability

Government/Public Sector Organizations
- Provision of services

- Fulfillment of social/legal mandate

Although there is an underlying commonality of attributes and elements of a quality system for these three organization types, it is still imperative to develop a suitable quality model commensurate with the specific needs of the enterprise.

The question that confronts most managers is: how to select or develop the most appropriate quality model. The marketplace abounds in recipes, models, theories, and experts on the subject of quality improvement. The problem, therefore, is not one of scarcity but of abundance and selecting the most appropriate strategy and model. A TQM model can be developed and established in a variety of ways. Following are some of the possible options:

- Develop your own system based on the fundamental principles of quality.

- Follow the guidelines recommended in the International Standard, ISO 9004-1.

- Follow the philosophies and methodologies propounded by quality experts such as Deming, Juran, Crosby, Feigenbaum, and Ishikawa.

Although any approach or model can provide a suitable quality system framework, the greatest lasting success is achieved by those companies and organizations who develop their own quality model that fits their own specific needs and infrastructural profile. A self-developed/self-directed model has a high probability of success because it is:

- Developed by people within the organization

- Compatible with the existing quality framework of the organization

- Making optimal use of available resources

- Participative, meaningful, value-adding and understandable to everyone collectively

- People-developed/people-empowered/people-driven/people-owned

TQM IMPLEMENTATION APPROACH

There are two fundamental approaches to implementing TQM in any organization; the top-down approach or the bottom-up approach.

Top-Down Approach: This exercise starts from the top—the Chief Executive Officer (CEO). The whole organization makes a decision to implement TQM and, accordingly, the development of strategic directions, decisions, and activities permeate throughout the organization top-down with everyone involved in the implementation process. The flow of activities may follow this sequence:

- The executive management identifies the need for improving quality and outlines their commitment, support, and involvement.

- A suitable quality infrastructure is established: TQM coordinator, steering committee, organizational responsibilities, requisite personnel and financial resources, etc.

- A strategic plan and a viable road map is drawn up.

- A suitable quality model and implementation plan is developed.

- Organizational policies, objectives, and goals are outlined.

- Quality improvement teams are established.

- Projects are initiated and implemented.

- Results are measured and analyzed.

- Quality system effectiveness and performance is evaluated.

- Continuous improvement opportunities are further explored and acted upon.

Bottom-Up Approach: This approach is applicable when there is no companywide decision on TQM implementation but there is a silent quality revolution brewing and everyone or several components in the organization are interested in improving the quality of their products, services, or processes. The quality improvement initiatives can be started project-by-project or process-by-process and ultimately the collective impact permeates into the organization upwards to bring about a total quality system implementation. The process can start from one or several divisions in the organization. The sequence of activities can be outlined as follows:

- Identify all the processes within a division where quality/process improvement efforts can be expended.

- Establish process management teams (PMTs).

- Follow the process improvement methodology of Plan-Upgrade-Record-Improve.

- Interface and extend improvement initiatives to other processes within the division and across divisions in the organization.

- Coordinate improvement initiatives and activities.

- Measure results and assess performance.

- Integrate process improvement efforts to bring the activity to the level of companywide TQM.

Irrespective of which approach is being used, top-down or bottom-up, the emphasis should clearly be on proceeding slowly and systematically to achieve the ultimate goal—*excellence in quality*.

QUALITY DILEMMA: SHORT-TERM OR LONG-TERM

Different organizations react differently to quality improvement pressures. Those in the reactive mode have generally a short-term focus and, consequently, their efforts are limited to quality control and quality assurance activities to improve the quality of their products and services. Proactive organizations operate on longer-term goals. Their efforts are expended on improving the quality of all aspects of work-life. While ensuring continuous quality improvement of their products and services, these companies are also concerned about the welfare of their employees, total customer satisfaction, partnership with their suppliers and subcontractors, and service to society in general. By doing so, they virtually guarantee for themselves an expanded marketability, continuous growth, high productivity, profitability, and long-term survivability.

Quality is a long-term perspective—it takes time to build quality in. Quality is not an entity you can buy—it requires patience, dedication, and planning. Most corporate boardrooms understand and agree that quality improvement will take time; yet, they often still rush into implementing quick-fix programs and start expecting large-scale improvements overnight only to encounter profound failures and frustrations.

Despite the understanding that short-term actions will only produce short-term success, why do most companies jump into this morbid chasm? Because, as the argument goes, if we don't produce quality overnight, we won't survive the next morning and the long-term focus, therefore, becomes meaningless. But the incongruence is, there is no survival even with the short-term initiatives. Having achieved quality and then losing it is more profoundly perilous than never achieving it in the first place. Regretfully, the quality dilemma is much like politics. If the government and politicians of the day do not produce quick results within the short period of their tenure, they don't get re-elected. Yet if they do operate on short-term visions, the country suffers for it.

The solution to this dilemma is not as complicated as most organizations may think. What is required is embedding into the planning framework an optimal mix of both the long-term as well as the short-term focus. In setting up a simple strategic plan, properly set your priorities. Develop a *Must-List* and a *Wish-List* of the deliverables vis-à-vis your short-term and long-term perspectives and needs. Appropriately maneuver and allocate your resources to both lists. Then measure, analyze, and review performance, revising and resetting priorities when necessary. At all times maintain your focus on the prioritized short-term as well as long-term goals.

TQBT: TOTAL QUALITY BUSINESS TRANSFORMATION

In searching for a suitable TQM model, the foremost concern of the organization is to ensure that both the short-term and longer-term needs are accommodated. Organizations are cognizant of the fact that quality is a long-term perspective and they are willing to make sacrifices to a feasible extent. However, they have to live by the fact that there are short-term day-to-day market pressures and demands which must be fulfilled.

The company's longer-term quality perspective revolves around establishing a sustainable quality culture and a mind-set for excellence. The short-term needs call for establishing a quality system, a process improvement strategy, and motivating people to improve quality and productivity. Figure A-1 presents a schematic of a layered quality model addressing these needs.

TQBT SYSTEM

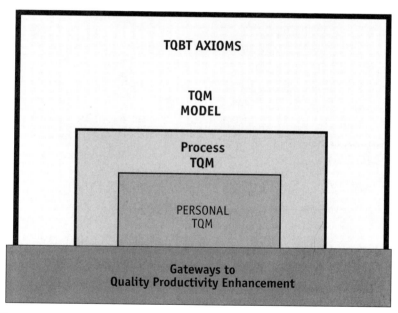

Figure A-1. TQBT System: Total Quality Business Transformation

The overall intent of any quality system, vis-à-vis the long-term and short-term focus, is to:

- Establish a sustainable quality culture

- Inculcate a mind-set for performance

- Improve quality of products and services

- Maintain an amicable relationship with suppliers and customers

- Achieve total customer satisfaction

- Improve market share and profitability

- Enhance productivity and reduce costs

- Create a pleasant and enjoyable working milieu for employees

The quality paradigm that we are presenting comprises a whole integrated system approach to quality. The model looks at the quality focus from the viewpoint of *Total Quality Business Transformation (TQBT)*. Depending upon the specific long-term and short-term needs of an enterprise, this layered quality model can be utilized either in its entirety or in parts. The message and approach inherent in the various components of the model, as schematically presented in Figure A-1, and Tables A-1, A-2, and A-3 is almost self-explanatory. The basic premise of the *TQBT System* has four important tiers:

1. **TQBT Axioms:** Implementing a quality culture

2. **TQM Model:** Implementing a total quality system

3. **Process TQM:** Process-by-process enhancement system

4. **Personal TQM:** A self-superimposed personal excellence mandate

In the next four sections we will discuss separately each tier of the TQBT System and provide guidelines for developing a workable quality model.

TQBT AXIOMS

The first tier of the system consists of twelve *TQBT Axioms* and serves as the umbrella for creating a long-term quality culture and high performance mind-set (see Table A-1). Indeed, the model requires management ingenuity, commitment, and dedication to develop means of permeating the message inherent in these axioms throughout the organization. Some basic guidelines for implementing the TQBT Axioms follow.

Table A-1. TQBT Axioms

Axiom #1:	Dynamic customer orientation
Axiom #2:	Participative management
Axiom #3:	Infrastructural stability
Axiom #4:	Self-directed organization
Axiom #5:	Operational framework flexibility
Axiom #6:	Integrated systems approach
Axiom #7:	System effectiveness measurability
Axiom #8:	Process enhancement focus
Axiom #9:	Genuine employee emphasis
Axiom #10:	Continuous reengineering
Axiom #11:	Proactive change management
Axiom #12:	Total excellence mind-set

TQBT Axioms: Implementation Guidelines

Dynamic Customer Orientation: Develop a mechanism through which a continuous linkage can be established with the customer, for example:

- Consultative process and meetings
- Feedback questionnaire/survey
- Customer hotline
- Personal visits

Pursue a proactive and dynamic customer satisfaction policy and perspective. Assure the customer, through some tangible means, that the organization has a business mandate, dedication and commitment to customer satisfaction. Make known to all employees how seriously the organization views the issue of customer satisfaction.

Participative Management: Management commitment alone is not enough; management must actively participate and involve itself in the process of building and maintaining the quality system. Some of the aspects of involvement may include: self-directed work teams, steering group meetings, system review meetings, strategic planning, and system auditing.

Infrastructural Stability: Organizational and/or infrastructural stability and maturity elevate morale and confidence. A stable and mature environment provides a pleasant working milieu. A perpetually changing infrastructure creates unpredictability and insecurity. It is, therefore, imperative to establish a sense of operational stability in the functional units. Even when an infrastructural change is necessary and unavoidable, make sure that the change is neither demeaning nor dehumanizing. People must fully participate in the change process.

Self-Directed Organization: As much as possible, the organization should be self-dependent and self-sufficient. Develop your own quality system/model, if possible, commensurate with your own needs and operating principles. A people-developed/people-driven/people-empowered/people-owned model would have the greatest chance of lasting success. Even if a model is adopted from external sources, make sure that the process ownership is in the hands of the employees. People must have the authority, empowerment, and responsibility to institute any necessary changes in the system to make it more efficient. Let not systems drive people; let people drive the systems.

Operational Framework Flexibility: Systems cannot afford to be inflexible in a volatile environment. Rigid systems which have little alignment with changing needs and priorities would soon become outmoded and obsolete. Continuous environmental scanning is required to acquire an up-to-date understanding of changing marketplace needs, demands, paradigm shifts, and customer requirements. Your internal systems must be flexible enough to admit and accommodate these changing and evolving demands.

Integrated Systems Approach: The functional framework of today's business world is complex and intricate. Its management calls for establishing an integrated cross-functional network in the organization. A coordinated whole system approach ensures operational consistency and uniformity. An effective cross-functional interface can be developed through self-directed work teams. Effective information systems and operational transparency generates a feeling of self-confidence in employees and as a result there is a high level of motivation that fuels productivity improvement.

System Effectiveness Measurability: Subjective assessment of system effectiveness or the extent of accomplishments has limited value. It is fundamentally important to establish viable measurement systems to provide solid data and information that can be properly analyzed through analytical means. No organization can afford to continue building a system without appraising its viability and usefulness. A step-by-step measuring and monitoring system should be an integral part of any quality system.

Process Enhancement Focus: Quality cannot be inspected-in. Final product inspection can only identify the product as conforming or nonconforming—it canot improve the product. The final product is simply a sum total of several processes spread over the life cycle of the product. Quality improvement starts at the process level and when improvement efforts are used at this level there is a greater likelihood of product conformance. To expect to achieve higher levels of product quality, therefore, requires continuous monitoring, control, and improvement of process. A simple, but highly effective, step-by-step process enhancement model is presented in the section, Process TQM.

Genuine Employee Emphasis: The success of any undertaking in the organization is dependent on its people. Today's workplace is highly democratized and we are back to the realization that human endeavor and commitment is singularly the most indispensable driving force for the success of any system. It is, therefore, imperative to focus our attention on creating the most conducive working environment for our people. People are at their best when they are given responsibility, are empowered to make decisions, are well informed, and the organization demonstrates a genuine concern and appreciation for their contribution. Employee participation and productivity can be significantly enhanced through the establishment of process management teams.

Continuous Reengineering: Dormancy breeds decay. To succeed, organizations must constantly reengineer, evolve, and adapt their strategies and systems to the demands of marketplace. This TQBT axiom emphasizes the need for continuous:

- Environmental scanning
- Customer needs analysis
- Organizational capability assessment
- System evaluation
- Employee emphasis
- Improvement focus

Proactive Change Management: Change is inevitable. It is indispensable for avoiding stagnation and remaining viable in the marketplace. However, change cannot be bought about by dictate. Change is a people-dependent phenomenon and for change to be accepted wholeheartedly, people must perceive a congruence between the nature of change and their beliefs. Managers must act as change agents. The management must provide appropriate support systems to make the process of change as smooth as possible.

Total Excellence Mind-set: The final TQBT axiom accentuates the fact that quality is a thought revolution in management—a collective attitude of mind. Quality means innate excellence. Quality does not come through implementing just a few programs or projects; it requires a vision, a dedication to a cause one truly believes in, a total excellence mind-set.

TQM MODEL

Although the initiatives and activities for a long-term quality focus as outlined in the TQBT Axioms are imperative, there is always that need of an organization to implement a quality system that can be used instantly, on a day-to-day basis, to improve the quality of products and services and to satisfy customers' needs and expectations. This is achieved through the second tier of the TQBT System, the *TQM Model*, which also becomes the basis for achieving quality system certification.

A typical TQM Model, irrespective of any approach or philosophy, would at least have these three following components:

1. Management structures and processes

2. Quality system procedures and processes

3. Disciplined tools and methodologies

Management Structures and Processes
- Sets the stage for establishing a quality culture and high performance mind-set

- Identifies management commitment and leadership

- Formalizes quality policies, objectives, responsibilities, and authority

- Provides structure and regimentation to all requisite systems

Quality System Procedures and Processes
- Establishes all requisite quality system procedures

- Provides a disciplined and structured systems approach to quality improvement

- Helps to establish formal standard operating procedures

- Facilitates proper documentation of procedures

Disciplined Tools and Methodologies
- Helps to measure and analyze process improvement and system effectiveness

- Assists in standardizing the processes and monitoring their continued viability

- Assists in continuous evaluation and improvement of the quality system

Some of the essential elements in these three components of a *TQM Model* would include the following:

Management Responsibilities

Vision	Employee Involvement
Mission	Teamwork
Commitment	Support Systems
Responsibility	Disciplined Methodology
Cultural Change	Knowledge and Skills

Leadership   Customer Focus

Support Systems

Procurement Control   Evaluation

Design Control   Quality Audits

Production Control   Verification Control

Process Control   Quality Records

Inspection and Testing   Training

Nonconformity   Servicing

Corrective and Preventive Action   Marketing Control

Cost Control   Post-Production Control

Documentation   Customer Feedback

Tools and Methodologies

Brainstorming   Quality Function
                                    Deployment

Flow Chart   Force Field Analysis

Checksheet   Shewhart-Deming Cycle

The Visual Factory   Nominal Group Technique

Cause-Effect Diagram   Benchmarking

Pareto Chart   Block Diagram

Just-in-Time   Relations Diagram

'PURI' Process Enhancement   Affinity Diagram/KJ
Wheel   Method

Control Charts   Matrix Diagram

Statistical Process Control	Matrix Data-Analysis
Design of Experiments	Process Decision Program
Systematic/Tree Diagram	Chart (PDPC)
Arrow Diagram	Concurrent Engineering

From the system elements listed above, we can now develop a cus-tomized self-developed, sustainable TQM Model. Under the overall umbrella of TQBT, Table A-2 presents the Ten TQM Absolutes: a ten-step plan for implementing the fundamental TQM elements at the operational level. A simple road map for implementing the *Ten TQM Absolutes* follows the table.

Table A-2. TQM Model: The Ten Absolutes

1. Management Readiness

2. Customer-Supplier Partnering

3. Environmental Scanning

4. Current System Evaluation

5. Strategic Planning

6. TQM Training

7. Disciplined System Implementation

8. Process Enhancement

9. Performance Evaluation

10. Continuous Improvement

TQM Model: Implementation Road Map

Step 1: Management Readiness

- Communicate management commitment to all levels of the organization.

- Arrange for a management orientation and awareness session. Seek the assistance of an outside consultant if required.

- Develop entities and practices to identify and highlight management commitment and participation.

- Management should assure the external customers and internal employees of their long-term commitment to a sustainable quality management system.

- Establish a steering committee/council to oversee and steer the quality efforts.

- Identify a TQM coordinator responsible for the day-to-day activities relating to system implementation, monitoring, performance evaluation, review, and continuous improvement.

- Develop vision and mission statements.

- Formulate quality policy and core values commensurate with the organization's goals and the expectations and needs of its customers.

Step 2: Customer-Supplier Partnering

- Highlight activities to identify dynamic customer focus.

- Develop a profile of customers' needs and expectations.

- Establish a customer service section to seek continuous feedback from the customers, buyers, or ultimate consumer.

- Develop a customer satisfaction profile.

- Identify and work with a select number of suppliers who are capable of providing high quality materials, services, or subcontracting services.

- Establish a close working relationship with suppliers. Make sure that the suppliers clearly understand the company's requirements. Assist the suppliers in monitoring and controlling the quality of their materials, products, and services.

- Establish practices to highlight constancy and trust with both the suppliers and the customers.

Step 3: Environmental Scanning

- Develop a profile of the market needs and demands in relation to your product or service offerings.

- Identify what your clients expect in terms of quality, technological innovativeness, and price.

- Identify other new product designs penetrating the marketplace.

- Assess your own market share. Compare your market share with that of your competitors. Identify opportunities for growth and improvement.

- Implement a continuous improvement strategy for developing new and innovative product lines for your potential customers and future markets.

- Identify needs, requirements, and expectations of your employees and establish suitable programs to enhance the quality of work-life.

Step 4: Current System Evaluation

- Conduct a gap analysis to evaluate strengths and weaknesses of your current system in relation to the needs and expectations identified through environmental scanning.

- Evaluate current systems in terms of their:

  – suitability of infrastructure for quality improvement

  – adequacy of requisite resources

  – adequacy of responsibility structure

  – adequacy of procedures, processes, and methodologies

  – satisfying the internal and external customer needs and demands

  – compliance with the regulations and codes of conduct

- Identify the status of system documentation and evaluate their operational suitability and adequacy.

- Identify the suitability of training and development activities.

Step 5: Strategic Planning

- Establish and document quality objectives, targets, and goals.

- Communicate quality objectives to all levels of the organization.

- Select or develop an appropriate TQM Model. The model must:

  – be commensurate with the current systems, procedures, and practices

  – be user–friendly

- involve people at all levels in the organization

- be sustainable

• Develop a master TQM implementation plan with time schedules, responsibilities, checkpoints, review procedures, performance evaluation framework, and activities for continuous improvement.

• Identify and allocate appropriate resources: human, financial, technological.

• Assign appropriate responsibilities and authority. Establish accountability framework. Document and communicate responsibilities and authority to everyone in the organization.

• Establish teams: process management teams, process improvement teams.

• Involve and empower people at all levels of the organization to implement, maintain, and improve the quality system.

• Develop and maintain quality system documentation hierarchy: quality manual, quality system procedures, standard operating procedures, work instructions, etc.

Step 6: TQM Training

• Establish a TQM training schedule for all employees. A typical sequence of training activities would include the following:

  - a half–day awareness session for executive management

  - a one to two day session on TQM implementation for all key managers and supervisors

  - several short sessions on TQM Model requirements for employees at all levels of the organization

- Establish training schedule for key personnel on:

  - process management and process improvement methodologies

  - developing, maintaining, and managing of operational procedures

  - quality system documentation

  - quality system auditing

  - process capability analysis

Step 7: Disciplined System Implementation

- Establish, maintain, and utilize documented procedures (standard operating procedures) in all areas of activity. These procedures must be:

  - available at all locations as required

  - accessible by everyone who needs them

  - reviewed and revised regularly

  - properly controlled

  - managed by designated responsibility centers and contact points

- Establish control points at all areas of activity, such as: procurement control, process control, production control, design control, and customer service.

- Establish a master control system for all documents and data pertaining to the quality system.

- Monitor interactive aspects of system implementation, such as: identification of nonconformities, internal quality system auditing, taking of corrective and preventive action, system reviews, and management reviews.

- Establish and maintain training and development programs.

- Establish employee motivational programs.

- Establish system audit and review schedules.

- Establish mechanism for identifying new initiatives and development projects.

- Establish mechanisms for cross–functional interaction and interfacing of process management teams.

- Establish a suitable quality records system.

Step 8: Process Enhancement

- Establish procedures to measure, monitor, and analyze key quality characteristics at all process control points.

- Establish procedures for the review and revision of standard operating procedures and other quality related procedures.

- Establish process improvement teams to evaluate process outputs and to recommend solutions and initiatives for process enhancement.

- Make use of suitable process improvement tools and statistical process control methods to improve processes.

- Establish appropriate automated and computerized systems to monitor and control the processes.

- Develop suitable work instructions for operational processes and utilize these for continuous training and retraining.

Step 9: Performance Evaluation

- Establish procedures for conducting system performance reviews at defined intervals.

- Establish procedures for the identification of nonconformances and taking of corrective and preventive action of each operational unit.

- Conduct regular system audits.

- Quantify and document performance evaluation review findings.

- Identify and document the corrective and preventive actions to be taken as a result of the evaluations.

- Document improvements made.

- Revise operating procedures in relation to the corrective and preventive actions.

Step 10: Continuous Improvement

- Establish a tangibly identifiable program of continuous improvement.

- Identify improvement opportunities.

- Establish strategic initiatives and allocate requisite resources.

- Develop a continuous training and development program.

- Develop employee empowerment, motivation, and incentive programs.

- Continuously partner with suppliers and customers.

- Focus on never–ending cycle of continuous improvement.

PROCESS TQM

Process TQM is the third and the most important tier of the TQBT System. Continuous process improvement is an integral part of any quality system. To improve the system, it is imperative to improve the individual processes. For any process improvement exercise, it is important for the team to develop a clear sense of direction with regard to the following:

- What needs to be improved?

- What is the current best?

- What is the goal?

- How would the goal be achieved?

- How would the achievement be measured?

The "PURI" Process Enhancement Wheel, as presented in Figure A–2, provides simple cyclic guidelines for continuous process quality improvement. A simple sequence of steps for process improvement follows the figure.

Process TQM: Procedure and Targets

- Identify the characteristics, variables, and attributes of the process that need to be improved.

- Identify the state of its current level of excellence.

- Define a goal for enhancing each or all of the characteristics.

- Establish a strategy for implementing the improvement effort.

- Identify improvement initiatives and projects.
- Implement the improvement plan.
- Monitor and control the improvement efforts.
- Identify deficiencies in the improvement initiatives.
- Take corrective and preventive action.
- Continue monitoring the performance.
- Evaluate results.
- Identify the improvements achieved.
- Standardize the achieved improvements.
- Adjust the process to the new standards and aims.
- Continue the cyclic process of improvement.

Figure A-2. "Puri" Process Enhancement Wheel

PERSONAL TQM

The fourth and last tier of the TQBT System pertains to a rather new and unique concept that we are calling *Personal TQM.*

An organization is merely a sum total of individuals. The success of any activity or undertaking in the organization is, indeed, in direct proportion to the nature and extent of the contributions made by its members, collectively as well as individually. To make the continuous quality improvement exercise a total success, diligence of action is, therefore, required from both ends: organizational thrust and individual impetus. TQM success is people–dependent.

It is my belief that it is the responsibility of the organization to provide a suitable operational environment, but it is an individual's own dedication and commitment that are key to success in any undertaking. The organization bears profound responsibility for rejuvenating and reenergizing the workplace. The extent to which individuals can exploit and utilize their potential depends heavily on the receptiveness of the working environment. The organization has to endeavor to reengineer the environment so that it is conducive to the optimal utilization of an individual's capabilities. However, no amount of organizational momentum can generate sustainable quality improvement impetus unless it is also genuinely supported and shared by the enthusiasm of the individual workers.

The organization can play its role in a variety of ways by proactively providing the requisite support. Some of the critical areas requiring special attention are:

- Identification of individual's strengths and weaknesses

- Identification of worker's needs: operational, functional, personal

- Identification of appropriate opportunities

- Development of suitable initiatives

- Provision of adequate resources

- Provision of effective leadership and direction

- Development of team approach

- Empowerment of people

- Effective change management

Personal TQM refers to the individual's own desires and efforts to improve performance and contributions to him- or herself, the organization, and to the society at large. It is a sense of self–imposed accountability, regimentation and a genuine desire to improve quality and performance, the resultant effect of which would be:

- Self–fulfillment, embellishment, and satisfaction

- Organizational growth and enhancement

- Societal enrichment

The role of an organization can go as far as establishing viable systems and creating a suitable working environment to encourage people to achieve lasting success. The focus, of course, must be on the people rather than the system. You don't drive the systems—you have to drive the people who would then drive the system. But the irony is that if people are not self–motivated and self–driven, no amount of organizational effort can bring fruit. As the saying goes, "You can lead a horse to water, but you can't make him drink."

In this section, we are striving to make a bold attempt to appeal to the psyche of every individual worker to rise above the ordinary, to set goals for themselves, and to give their personal best, individually and collectively, towards a shared vision of excellence. Table A–3 sets out the basic ten commandments of Personal TQM.

Personal TQM has very little to do with any organizational performance appraisal or rating system. It is a proactive approach to self–awakening and self–fulfillment. To generate a framework for Personal TQM, the following sequence of actions is suggested:

Introspective Analysis: identify your current state of functional performance

Gap Analysis: identify your strengths and weaknesses

Needs Analysis: identify what needs to be done to enhance your performance

Action Plan: develop a systematic approach to implement the planned set of activities

Performance Analysis: evaluate the extent of improvements made

Table A-3. Personal TQM: The Ten Commandments

1. Create a vision

2. Understand personal role

3. Determine personal potential

4. Respond to challenge

5. Practice participative involvement

6. Cultivate proactive attitude

7. Instill creative mindedness

8. Impose self-directed performance appraisal

9. Assess personal fulfilment

10. Exercise continuous introspective analysis

As a first step to Personal TQM, every individual should undergo an introspective analysis and identify his strengths and weaknesses, needs and requirements, personal and organizational contributions, and suitable framework for improvement. Following are some of the critical factors that needs to be probed in an introspective analysis.

Personal TQM: Introspective Analysis

- What is my role in the organization?

- In what processes am I involved?

- Do I know how and where my role and work fits in the overall organizational framework?

- Do I have the requisite resources to do my job?

- Do I have proper education, training, and experience to do my job?

- Who is my inter/intra cross–functional interface?

- How helpful am I to others?

- Do I have a long–term plan to improve the quality of my work?

- Am I managing my time well?

- Am I doing my job productively?

- At the end of the day, can I tangibly identify my accomplishments?

- Have I ever analytically measured and evaluated my performance, productivity, and effectiveness?

- Am I limited to doing what someone tells me or am I a self–motivated innovator?

- Am I empowered?

- Do I take control and ownership of processes in my own hands?

- Do I operate in an active or proactive mode?
- How do I work in a team environment?
- Do I actively and effectively participate in a team effort?
- Do I willingly accept responsibility?
- How do I handle change?
- What is my overall contribution to the organization?
- What is my contribution to myself?
- What is my contribution to society?

An introspective analysis such as this should provide sufficient information to proceed further to identify:

- Personal strengths and weaknesses
- Personal needs and requirements
- The nature and extent of organizational support required
- Basic framework for improvement

From here onwards, it is a joint responsibility of the management and the individual: the management to provide effective support systems for the individual's development, and the individual to work towards self–improvement and genuine concern for the welfare of the organization.

TQM: THE FINALE

Summarizing the above discussion on TQM, the following order of events is recommended for the implementation of a TQM model:

- Top management undergoes TQM orientation.

- Top management and senior executives become involved and lead the way.

- Management identifies commitment, makes quality the top priority, and establishes a strategic quality plan.

- The company seeks customer involvement and input.

- Employees at all levels are trained in appropriate aspects of TQM.

- Employee involvement is sought through the establishment of a TQM Steering Committee and Process Management Teams (PMTs).

- Qualified suppliers are identified and a quality partnership is established.

- Management develops a quality policy and corresponding objectives.

- A cross–functional quality infrastructure is established with appropriate delineation of responsibilities.

- TQM elements, procedures, and processes are developed and adopted on a consistent basis.

- Quality improvement projects and initiatives are implemented.

- Data and information are collected and analyzed to study the effectiveness of the TQM Model.

- System deficiencies are corrected and preventive measures are implemented.

- Quality improvement results are demonstrated.

- A continuous cycle of improvement is followed.

In order to ensure effective TQM implementation and sustained quality improvement levels, a complete cultural transformation is essential. Following are some of the key factors to be considered:

- Ensure that management is truly committed and demonstrates its commitment across the entire organization.

- Develop open, responsive, group–driven quality leadership.

- Continuously reassert that quality is everyone's responsibility and not just that of a few key people. There is a need to be obsessed with quality and excellence.

- Strike a balance between long–term goals and successive short–term objectives. Establish a sense of constancy of purpose and a focus on long–term continuous improvement.

- Use a systematic approach and disciplined methodology to clearly understand external and internal customer requirements. Make the organization customer–satisfaction driven.

- Establish a mutually supportive control and improvement partnership with the suppliers.

- Use process management teams to involve everyone and seek improvement ideas and opportunities.

- Recognize employee achievement and establish an effective incentive program.

- Institute a continuous process of education, training, learning, and self–improvement.

- Ensure that the emphasis on customer focus and continuous improvement permeates the whole organization.

Finally it must be emphasized that maintainability and sustainability of the system is of paramount importance. A TQM exercise is not an entity with a limited shelf–life; it is an exercise on a continuum, one which has no beginnings or endings. It is a race without the start or finish line. One cannot simply and abruptly start a full scale TQM program in a day and end the program at a predetermined stipulated date. The system requires time and patience to establish and it must be sustainable. A word of advice—it would be better for an enterprise not to venture into implementing a quality system in the first place if it cannot maintain it. It may be possible to survive without a system, but having established one and losing it is dangerous. The justification for not having a quality system may be plausible, but the loss of credibility upon failure of the system can be unrecoverable.

The key characteristics of a sustainable TQM, therefore, are:

- The system keeps on going and no one thinks that TQM came and went.

- The quality system becomes second nature to everyone.

- The system has a built–in self–correcting, self–rejuvenating, and self–reenergizing mechanism.

- The system is people–driven and people empowered.

- The system belongs to everyone and not just a few people in the organization.

- The management continuously supports the system and commits adequate resources.

- The system does not get overburdened by unnecessary documentation and procedural stringency, complexity, or inflexibility.

- The system is flexible and controllable.

- The system is simple and user–friendly.

- The system is adequately linked to new and innovative ideas and continuous improvement initiatives.

- The system makes everyone a winner.

BIBLIOGRAPHY

Puri, Subhash C. *ISO 9000 Certification and Total Quality Management, 2nd ed.* Nepean-Ottawa: Standards-Quality Management Group, 1995.

Puri, Subhash C. *Statistical Process Quality Control — Key to Productivity.* Nepean-Ottawa: Standards-Quality Management Group, 1984.

Puri, Subhash C. *TQM/ISO 9000/SPC: Why Do Systems Fail.* Transactions 49th Annual Quality Congress. Cincinnati: American Society for Quality Control, 1995.

Canadian Standard, Z750: A Voluntary Environmental Management System. Rexdale (Toronto): Canadian Standards Association, 1994.

British Standard, BS 7750: Environmental Management Systems. London: British Standards Institution, 1994.

South African Standard, SABS 0251: Environmental Management Systems. Pretoria: South African Bureau of Standards, 1993.

International Organization for Standardization (ISO). *ISO 14001: Environmental Management Systems - Specifications with Guidance for Use.* Geneva: International Organization for Standardization, 1996.

ABOUT THE AUTHOR

Subhash C. Puri is an internationally renowned author, lecturer, and consultant on the subject of quality. As principal of Standards-Quality Management Group, an Ottawa (Canada) based consulting company, he provides training and consultation on a variety of subjects such as: TQM, ISO 9000 and ISO 14000 Certifications, Quality System Auditing, SPC, Business Process Reengineering, etc. Prior to starting his own consulting and training activities, he served as Director and Chief Statistician at Agriculture Canada and taught at several universities in Canada and abroad.

As one of the leading authorities on the subject, he has been a keynote speaker and lecturer for many associations and organizations in several countries. For over two decades, he has provided extensive training and consultation to numerous organizations and companies in the manufacturing, service, and public sectors at the national and international levels, for implementing TQM and ISO 9000 systems.

Being actively involved in the national and international standardization activities, he has made significant contributions to the development of statistical and quality standards. He is the chairman of CAC/ISO/TC69, member of CAC/ISO/TC176 and has served as chairman/member on many other standards committees. He is the author of several books and has published numerous professional papers on the subject.

OTHER TITLES BY THE AUTHOR

Puri, S.C. and Mullen, K. *Applied Statistics for Food and Agricultural Scientists*. Boston, Massachusetts: G.K. Hall & Co., 1980.

Puri, S.C., Ennis, D., and Mullen, K. *Statistical Quality Control for Food and Agricultural Scientists*. Boston, Massachusetts: G.K. Hall & Co., 1979.

Puri, S.C. *Statistical Methods for Food Quality Management*. Ottowa, Canada: Agriculture Canada, Publication number A73-5268, 1989.

Puri, S.C. *Statistical Aspects of Food Quality Assurance*. Ottowa, Canada: Agriculture Canada, Publication number 5140, 1981.

INDEX

Books from Productivity Press

Productivity Press publishes books that empower individuals and companies to achieve excellence in quality, productivity, and the creative involvement of all employees. Through steadfast efforts to support the vision and strategy of continuous improvement, Productivity Press delivers today's leading-edge tools and techniques gathered directly from industrial leaders around the world. Call toll-free 1-800-394-6868 for our free catalog.

16 Point Strategy for Productivity and Total Quality Control
William F. Christopher and Carl G. Thor

Major breakthroughs in productivity improvement can only be achieved when one is willing to make major changes. This book provides the definitive list of what must be considered when implementing continuous improvement methods throughout an organization.
ISBN 1-56327-072-2 / 69 pages / $15.95 / Order Item # MS7-B270

20 Keys to Workplace Improvement (Revised Edition)
Iwao Kobayashi

The 20 Keys system does more than just bring together twenty of the world's top manufacturing improvement approaches—it integrates these individual methods into a closely interrelated system for revolutionizing every aspect of your manufacturing organization. This revised edition of Kobayashiís best-seller amplifies the synergistic power of raising the levels of all these critical areas simultaneously. The new edition presents upgraded criteria for the five-level scoring system in most of the 20 Keys, supporting your progress toward becoming not only best in your industry but best in the world. New material and an updated layout throughout assist managers in implementing this comprehensive approach. In addition, valuable case studies describe how Morioka Seiko (Japan) advanced in Key 18 (use of microprocessors) and how Windfall Products (Pennsylvania) adapted the 20 Keys to its situation with good results.
ISBN 1-56327-109-5/ 312 pages / $50.00 / Order 20KREV-B270

The Benchmarking Management Guide
American Productivity & Quality Center

If you're planning, organizing, or actually undertaking a benchmarking program, you need the most authoritative source of information to help you get started and to manage the process all the way through. Written expressly for managers of benchmarking projects by the APQC's renowned International Benchmarking Clearinghouse, this guide provides exclusive information from members who have already paved the way. It includes information on training courses and ways to apply Baldrige, Deming, and ISO 9000 criteria for internal assessment, and has a complete bibliography of benchmarking literature.
ISBN 1-56327-045-5 / 260 pages / $40.00 / Order BMG-B270

Productivity Press, Dept. BK, P.O. Box 13390, Portland, OR 97213-0390
Telephone: 1-800-394-6868 Fax: 1-800-394-6286

Caught in the Middle
A Leadership Guide for Partnership in the Workplace
Rick Maurer

Managers today are caught between old skills and new expectations. You're expected not only to improve quality and services, but also to get staff more involved. This stimulating book provides the inspiration and know-how to achieve these goals as it brings to light the rewards of establishing a real partnership with your staff. Includes self-assessment questionnaires.
ISBN 1-56327-004-8 / 258 pages / $30.00 / Order CAUGHT-B270

Corporate Diagnosis
Meeting Global Standards for Excellence
Thomas L. Jackson with Constance E. Dyer

All too often, strategic planning neglects an essential first step-and final step-diagnosis of the organization's current state. What's required is a systematic review of the critical factors in organizational learning and growth, factors that require monitoring, measurement, and management to ensure that your company competes successfully. This executive workbook provides a step-by-step method for diagnosing an organization's strategic health and measuring its overall competitiveness against world class standards. With checklists, charts, and detailed explanations, *Corporate Diagnosis* is a practical instruction manual. The pillars of Jackson's diagnostic system are strategy, structure, and capability. Detailed diagnostic questions in each area are provided as guidelines for developing your own self-assessment survey.
ISBN 1-56327-086-2 / 100 pages / $65.00 / Order CDIAG-B270

Cost Reduction Systems
Target Costing and Kaizen Costing
Yasuhiro Monden

Yasuhiro Monden provides a solid framework for implementing two powerful cost reduction systems that have revolutionized Japanese manufacturing management: target costing and kaizen costing. Target costing is a cross-functional system used during the development and design stage for new products. Kaizen costing focuses on cost reduction activities for existing products throughout their life cycles, drawing on approaches such as value analysis. Used together, target costing and kaizen costing form a complete cost reduction system that can be applied from the product's conception to the end of its life cycle. These methods are applicable to both discrete manufacturing and process industries.
ISBN 1-56327-068-4 / 400 pages / $50.00 / Order CRS-B270

Productivity Press, Dept. BK, P.O. Box 13390, Portland, OR 97213-0390
Telephone: 1-800-394-6868 Fax: 1-800-394-6286

Handbook for Productivity Measurement and Improvement

William F. Christopher and Carl G. Thor, eds.

An unparalleled resource! In over 100 chapters, nearly 80 frontrunners in the quality movement reveal the evolving theory and specific practices of worldclass organizations. Spanning a wide variety of industries and business sectors, they discuss quality and productivity in manufacturing, service industries, profit centers, administration, nonprofit and government institutions, health care and education. Contributors include Robert C. Camp, Peter F. Drucker, Jay W. Forrester, Joseph M. Juran, Robert S. Kaplan, John W. Kendrick, Yasuhiro Monden, and Lester C. Thurow. Comprehensive in scope and organized for easy reference, this compendium belongs in every company and academic institution concerned with business and industrial viability.
ISBN 1-56327-007-2 / 1344 pages / $90.00 / Order HPM-B270

Implementing a Lean Management System

Thomas L. Jackson with Karen R. Jones

Does your company think and act ahead of technological change, ahead of the customer, and ahead of the competition? Thinking strategically requires a company to face these questions with a clear future image of itself. *Implementing a Lean Management System* lays out a comprehensive management system for aligning the firmís vision of the future with market realities. Based on Hoshin management, the Japanese strategic planning method used by top managers for driving TQM throughout an organization, Lean Management is about deploying vision, strategy, and policy to all levels of daily activity. It is an eminently practical methodology emerging out of the implementation of continuous improvement methods and employee involvement. The key tools of this book builds on the knowledge of the worker, multi-skilling, and an understanding of the role and responsibilities of the new lean manufacturer.
ISBN 1-56327-085-4 / 150 pages / $65.00 / Order ILMS-B270

Integrated Cost Management

A Companywide Prescription for Higher Profits and Lower Costs

Michiharu Sakurai

To survive and grow, leading-edge companies around the world recognize the need for new management accounting systems suited for todayís advanced manufacturing technology. Accountants must become interdisciplinary to cope with increasing cross-functionality, flexibility, and responsiveness. This book provides an analysis of current best practices in management accounting in the U.S. and Japan. It covers critical issues and specific methods related to factory automation and computer integrated manufacturing (CIM), including target costing, overhead management, Activity-Based Management (ABM), and the cost management of software development. Sakuraiís brilliant analysis lays the foundation for a more sophisticated understanding of the true value that management accounting holds in every aspect of your company.
ISBN 1-56327-054-4 / 300 pages / $50.00 / Item # ICM-B270

Productivity Press, Dept. BK, P.O. Box 13390, Portland, OR 97213-0390
Telephone: 1-800-394-6868 Fax: 1-800-394-6286

Learning Organizations
Developing Cultures for Tomorrow's Workplace
Sarita Chawla and John Renesch, Editors

The ability to learn faster than your competition may be the only sustainable competitive advantage! A learning organization is one where people continually expand their capacity to create results they truly desire, where new and expansive patterns of thinking are nurtured, where collective aspiration is set free, and where people are continually learning how to learn together. This compilation of 34 powerful essays, written by recognized experts worldwide, is rich in concept and theory as well as application and example.

An inspiring followup to Peter Senge's ground-breaking best-seller *The Fifth Discipline*, these essays are grouped in four sections that address all aspects of learning organizations: the guiding ideas behind systems thinking; the theories, methods, and processes for creating a learning organization; the infrastructure of the learning model; and arenas of practice.

ISBN 1-56327-110-9 / 575 pages / $35.00 / Order LEARN-B270

The TQM Paradigm: Key Ideas That Make it Work
Derm Barrett

For anyone beginning or already enmeshed in TQM implementation, this book is an excellent exploration into key concepts and the companywide changes in systems and culture that are necessary to manage the process.

ISBN 1-56327-073-0 / 74 pages / Order Item #MS8-B270 / $15.95 hardcover

Manufacturing Strategy
How to Formulate and Implement a Winning Plan
John Miltenburg

This book offers a stepbystep method for creating a strategic manufacturing plan. The key tool is a multidimensional worksheet that links the competitive analysis to manufacturing outputs, the seven basic production systems, the levels of capability and the levers for moving to a higher level. The author presents each element of the worksheet and shows you how to link them to create an integrated strategy and implementation plan. By identifying the appropriate production system for your business, you can determine what output you can expect from manufacturing, how to improve outputs, and how to change to more optimal production systems as your business needs changes. This is a valuable book for general managers, operations managers, engineering managers, marketing managers, comptrollers, consultants, and corporate staff in any manufacturing company

ISBN 1-56327-071-4 / 391 pages / $45.00 / Order MANST-B270

Productivity Press, Dept. BK, P.O. Box 13390, Portland, OR 97213-0390
Telephone: 1-800-394-6868 Fax: 1-800-394-6286

A New American TQM
Four Practical Revolutions in Management
Shoji Shiba, Alan Graham, and David Walden

For TQM to succeed in America, you need to create an American-style "learning organi-zation" with the full commitment and understanding of senior managers and executives. Written expressly for this audience, *A New American TQM* offers a comprehensive and detailed explanation of TQM and how to implement it, based on courses taught at MIT's Sloan School of Management and the Center for Quality Management, a consortium of American companies. Full of case studies and amply illustrated, the book examines major quality tools and how they are being used by the most progressive American companies today.
ISBN 1-56327-032-3 / 606 pages / $50.00 / Order NATQM-B270

Performance Measurement for World Class Manufacturing
A Model for American Companies
Brian H. Maskell

If your company is adopting world class manufacturing techniques, you'll need new meth-ods of performance measurement to control production variables. In practical terms, this book describes the new methods of performance measurement and how they are used in a changing environment. For manufacturing managers as well as cost accountants, it pro-vides a theoretical foundation of these innovative methods supported by extensive practi-cal examples. The book specifically addresses performance measures for delivery, process time, production flexibility, quality, and finance.
ISBN 0-915299-99-2 / 448 pages / $55.00 / Order PERFM-B270

Software and the Agile Manufacturer
Computer Systems and World Class Manufacturing
Brian Maskell

The term "agile manufacturing" describes responsive, flexible manufacturing that can deliver better products, faster, at lower cost. This book is the first to address the critical question of how computerization can aid the transition. It shows how computer systems and software designed for individual departments or functions can be adapted to create a world class manufacturing environment that's integrated companywide. Case studies reveal the common characteristics companies have shared in the challenge to comput-erize and provide guidelines for companies just starting out. This is a non-technical, practical guide.
ISBN 1-56327-046-3 / 424 pages / $50.00 / Order SOFT-B270

Productivity Press, Dept. BK, P.O. Box 13390, Portland, OR 97213-0390
Telephone: 1-800-394-6868 Fax: 1-800-394-6286

TO ORDER: Write, phone, or fax Productivity Press, Dept. BK, P.O. Box 13390, Portland, OR 97213-0390, phone 1-800-394-6868, fax 1-800-394-6286. Send check or charge to your credit card (American Express, Visa, MasterCard accepted).

U.S. ORDERS: Add $5 shipping for first book, $2 each additional for UPS surface delivery. Add $5 for each AV program containing 1 or 2 tapes; add $12 for each AV program containing 3 or more tapes. We offer attractive quantity discounts for bulk purchases of individual titles; call for more information.

ORDER BY E-MAIL: Order 24 hours a day from anywhere in the world. Use either address:
To order: service@ppress.com
To view the online catalog and/or order: http://www.ppress.com/

QUANTITY DISCOUNTS: For information on quantity discounts, please contact our sales department.

INTERNATIONAL ORDERS: Write, phone, or fax for quote and indicate shipping method desired. For international callers, telephone number is 503-235-0600 and fax number is 503-235-0909. Prepayment in U.S. dollars must accompany your order (checks must be drawn on U.S. banks). When quote is returned with payment, your order will be shipped promptly by the method requested.

NOTE: Prices are in U.S. dollars and are subject to change without notice.